Access your online resources

99 Eco-Activities for Your Primary School is accompanied by a number of printable online materials, designed to ensure this resource best supports your professional needs. Do think before you print about how many copies are required and if print-outs can be shared.

Activate your online resources:

Go to www.routledge.com/cw/speechmark and click on the cover of this book.

Click the 'Sign in or Request Access' button and follow the instructions in order to access the resources.

99 ECO-ACTIVITIES FOR YOUR PRIMARY SCHOOL

This book is packed with bright ideas and practical projects for children aged 4–11 to raise environmental awareness and prompt discussion about climate change.

Encouraging children to take charge right from the start, the activities range from creating recycled kites, windsocks, and garden decorations, to upcycling old t-shirts, building minibeast hotels and designing campaigns to eliminate single-use plastics from school. Some can be completed outdoors and some indoors, with each page including photos of the activity in action, plus details of the resources required and steps needed. As well as the main activity, extension ideas are provided, so there is plenty to fill each session. The tried-and-tested activities are themed in three main areas:

- Eco-friendly practice
- Recycling and upcycling
- Connecting with the natural world

Whether you run an eco-club, a craft club or you simply want to facilitate activities on a sustainability theme with children in your class, this fantastic book will raise awareness of environmental issues in an engaging way – and many of the activities will save your school or setting money too.

Sarah Watkins is a Forest School leader on a journey of climate change action. An ex-primary teacher and Head of School, Sarah has led eco-clubs and supported schools to apply for eco accreditation. Before teaching, Sarah worked as an Arts Officer, and also managed projects for a national media charity, giving a platform to unheard voices. Sarah currently lectures at her local university and writes for various publications. She also presents nationally on ways to involve children positively in environmental conversations.

"Eco activities for the present, helping children to learn and be aware for their future – after all, our children need to be the 'movers and shakers' for tackling climate change. Designed for children, but fully tried and tested on them too (and they always say it as it is) so you're guaranteed to have excellent and inspirational environmental learning taking place as a result!"

Julian Thomas, *Headteacher of Georgeham Primary School, the UK's first 'plastic free' school*

"I absolutely LOVED this book, the activities are so simple and easy to follow, they make use of ordinary, everyday materials, they promote eco-conscious thinking, and each activity comes with a 'Why' – kids love to ask 'WHY?'! The most inspiring activity book I have come across by far! (My favourite activity is number 36!)"

Isobel Mary Champion, *Parenting Coach*

"This book is full of fantastic ideas to support and engage children of all ages and stages in becoming 'sustainability heroes' and making a difference to our world in the process. Each page is bursting with activities that are both engaging and easy for supporting adults to set up and manage. Each activity supports the development of critical thinking skills and empowers learners to see that they are never too small to make a difference. An amazing addition to any teacher, parent, or carer's wish list."

Ashley Forrester, *Principal Teacher of ASN, SEND Teacher*

"This is the book our school has been waiting for! Our pupils recognise themselves as Eco warriors and work in partnership with national organisations to counteract the effects of climate change but what are the things that we can do daily in school and at home that enhances our understanding of these issues? Full of simple and useful activities that not only develop deeper understanding but also fire curiosity and creativity. I can see our children, staff and families adapting so many of these wonderful ideas and adding to them too. Thank you just isn't enough for an eco small book that will make a huge impact."

Sharifa Lee, *Headteacher*

"You would be hard pushed to find a more practical and easy to access book to enhance your outdoor teaching and learning. Every teacher should buy a copy for inspiration on how to deliver wonderful outdoor learning. It is perfectly timed for the eco agenda schools are heavily investing in and the book is full of wonderful ideas that will support teaching, stir up the creative juices and really get teachers inspired about their curriculum. 100% a must buy."

Vivien Watson, *Headteacher and Local Leader of Education, Hookstone Chase Primary School*

99 ECO-ACTIVITIES FOR YOUR PRIMARY SCHOOL

ENGAGING IDEAS THAT PROMOTE ENVIRONMENTAL AWARENESS

Sarah Watkins

LONDON AND NEW YORK

Cover image: Sarah Watkins

First published 2023
by Routledge
4 Park Square, Milton Park, Abingdon, Oxon OX14 4RN

and by Routledge
605 Third Avenue, New York, NY 10158

Routledge is an imprint of the Taylor & Francis Group, an informa business

© 2023 Sarah Watkins

The right of Sarah Watkins to be identified as author of this work has been asserted in accordance with sections 77 and 78 of the Copyright, Designs and Patents Act 1988.

All rights reserved. The purchase of this copyright material confers the right on the purchasing institution to download pages which bear the companion website icon and a copyright line at the bottom of the page. No other parts of this book may be reprinted or reproduced or utilised in any form or by any electronic, mechanical, or other means, now known or hereafter invented, including photocopying and recording, or in any information storage or retrieval system, without permission in writing from the publishers.

Trademark notice: Product or corporate names may be trademarks or registered trademarks, and are used only for identification and explanation without intent to infringe.

British Library Cataloguing-in-Publication Data
A catalogue record for this book is available from the British Library

Library of Congress Cataloging-in-Publication Data
A catalog record has been requested for this book

ISBN: 978-1-032-12301-1 (hbk)
ISBN: 978-1-032-12302-8 (pbk)
ISBN: 978-1-003-22399-3 (ebk)

DOI: 10.4324/9781003223993

Typeset in DINPro
by Deanta Global Publishing Services, Chennai, India

Access the companion website: www.routledge.com/cw/speechmark

A small step is still a step forward. We walk together...

CONTENTS

Foreword by pupils at
Stretton Sugwas Church of
England Academy   x
Acknowledgements   xii

Introduction   1

**1 Eco friendly practice:
Sustainability heroes**   3

2 Recycling and upcycling   18

3 Connect with the natural world   90

Conclusion   187

FOREWORD BY PUPILS AT STRETTON SUGWAS CHURCH OF ENGLAND ACADEMY

Congratulations on buying this life-changing book. It shows you care and want to make the world a better place, a better place for children like us and the next generations to come.

The number of people who believe in being 'eco' surprises us because no matter how hard we try, there are other people who always manage to ruin it- either by being lazy or just not knowing. Lots of animals are endangered because of us not being kind and responsible with our actions.

We believe that being eco is very important for our planet because it's ours, yours and every other living creatures' home. But as humans, we can be very wasteful.

Billions of tonnes of rubbish ends up in the ocean each *year*. Billions of tonnes of rubbish are taken to landfill, where it's literally dumped. Enough is enough.

When the world was in lockdown and grown-ups weren't using their cars all the time, the world started to repair itself. **Change is possible if we all work together.** If everyone believed in helping the world, it wouldn't be like this. We should all take part in keeping our world a cleaner and safer place for all. Let nature breathe.

Let's think about bees. Each bee plays a part in the hive. If we all work together, we can achieve great things. By reading this book, you join a big team of people who all want to work together to give the planet chance to heal and flourish.

Imagine if all of us did our part to do just 10 minutes of litter picking. So much of the planet would be cleaner. Imagine if we recycled and re-used things instead of throwing them away.

This is where this book comes in. It's packed with amazing ideas on how to reuse items, as part of an Eco Club or other group.

Foreword by pupils at Stretton Sugwas Church of England Academy

So read on to find out some tips and tricks on how *you* can help our planet and inspire children like us to do the same.

Change is possible if we all work together

Yours sustainably,
Joscelyn, Susie and Erin
Stretton Sugwas Church of England Academy

ACKNOWLEDGEMENTS

Thank you to *you* for making a difference. Thank you to Julian Thomas, Tamara Pearson and Ashleigh Robertson.

Thank you to every school, setting and individual who trialled and photographed an activity:

Aga Khan Academy Maputo
Albany Infant and Nursery School
Anthony Hadfield
Antingham and Southrepps Primary School
Ashdown Primary School
Blossom Pre-school
Branston C of E Infant School
British International School of Kuala Lumpur
Brockwell Junior School
Burlington Junior School
Carrongrange High School
Castlewood Primary School
Chantry Community Academy
Claire Peach at Hens for Hire
Copmanthorpe Primary School
Drapers' Pyrgo Priory School
Earthtime for All
East End Primary School
Forest Foxes
Fruits of the Forest
Gunthorpe C of E Primary School
Highters Heath Nursery School
Jane Hewitt
Langton Primary School
Laurel Saunders
Ledbury Primary School
Leticia Cariño

Acknowledgements

Liff Primary School
Little Sutton Primary School
Moira Primary School
Mylnhurst School
Newchurch St. Nicholas
Penybont Primary School
Quarry Bank Primary School
Saighton Church of England Primary School
Saint Mary's RC Primary School, Oswaldtwistle
Sarah Clarke
Solent Infant School
St Elphin's CE VA Primary School
St Issey C of E Primary
St. Denis' Primary School
St. Peter's Catholic Primary School
Stephanie Barber
Stretton Sugwas Church of England Academy
The Arbor School, Dubai
The LimbBo foundation
The Pines School
The Prince of Wales School
The Village Prep School
Thomas Coram C of E School
Upton Meadows Primary School
Vicki Waud
Writtle Infant School

INTRODUCTION

Climate change is happening right now. The last few years have been the warmest on record. Ice sheets in Antarctica and Greenland have shrunk. Glaciers around the world are disappearing and the sea level is rising.

This can seem overwhelming, and increasing numbers of children are experiencing climate anxiety. It's easy to feel powerless which can lead to excessive worrying, fear and sadness.

Climate change is a complicated issue that no one can solve on their own. Talking about the situation can help: you might be surprised by what scares children about climate change. Giving children the chance to talk through their worries helps them feel less alone and the simple act of saying it aloud can reduce anxiety. Sharing books that cover the issues can be a useful starting point for discussion and there is a list of suitable books to add to your library in the next section of this book.

It's helpful to talk about the positive things that are happening around the world: scientists and researchers are working relentlessly on inspirational ways to address climate change.

Children need to feel they can personally make a positive change through small, achievable actions. Encouraging children to focus on what they can control rather than what they can't is a good strategy to reduce climate anxiety.

The activities in this book prompt children to make a difference through changing established ways of doing things; recycling or building a connection with the natural world. Designed for primary schools or other settings like nurseries or community groups, each activity is suitable for a group of mixed-age children (so would also be suitable for families) and only requires easily available resources. Once you have the resources ready, every activity is written in a way that is accessible to children so they can take charge right from the start.

Introduction

Books that cover eco issues

Over in the Ocean: In a Coral Reef by Marianne Berkes
Look What I Found in the Woods by Moira Butterfield and Jesús Verona
My First Heroes: Eco Warriors by Campbell Books
The Last Tree by Ingrid Chabbert and Guridi
Charlie and Lola: Look After Your Planet by Lauren Child
Leaf by Sandra Dieckmann
Earth Heroes by Lily Dyu and Amy Blackwell
City of Rust by Gemma Fowler
What a Waste: Rubbish, Recycling, and Protecting our Planet by Jess French
Grandpa's Garden by Stella Fry
Tidy by Emily Gravett
The Great Paper Caper by Oliver Jeffers
Dear Greenpeace by Simon James
The Sea: Exploring Our Blue Planet by Miranda Krestovnikoff and Jill Calder
Song of the Dolphin Boy by Elizabeth Laird
A Planet Full of Plastic by Neal Layton
The Happy Hedgerow by Elena Mannion and Erin Brown
Fantastically Great Women Who Saved the Planet by Kate Pankhurst
One Plastic Bag by Miranda Paul and Elizabeth Zunon
George Saves the World by Lunchtime by Jo Readman
Someone Swallowed Stanley by Sarah Roberts
The Lorax by Dr. Seuss
How to Save the World with a Chicken and an Egg by Emma Shevah
The Tale of the Whale by Karen Swann
The Rabbits by Shaun Tan
The Last Wild by Piers Torday
Greta and the Giants by Zoe Tucker
10 Things I Can Do to Help My World by Melanie Walsh
The Big Book of the Blue by Yuval Zommer

Health and safety

The activities in this book are designed to be carried out with groups of children aged 4–11 and I've aimed to reduce the level of risk because I recognise that sometimes one adult will be supervising a large group. I'm sure that any adult reading this book will put the safety of the children first and work to minimise any harmful risks. If there are particular risks linked to water, heat or tools, for example, I have indicated this in the activity description. Please use common sense.

1
ECO-FRIENDLY PRACTICE
Sustainability heroes

Of course, we need to continue to put pressure on world leaders and big businesses to act more sustainably and make a genuine commitment to the environment. However, we shouldn't disregard the impact of our everyday actions. Making small changes in the way we do things can not only help combat climate change but also change other people's perspectives.

We're all vulnerable to 'busyness,' getting things done quickly and moving onto the next task. It can be easy to fall into the trap of always doing things in a certain way. Why not stop and take note of all the processes that take place at your school or setting over the course of a day or a week? Carrying out an audit can highlight practices that are harmful to the environment and at this point, think big! You may come across what I call 'inflexibles,' people who will tell you that 'it's always been done this way and we can't change it.' Rip up the rulebook and be an innovator! (The planet will thank you.)

The activities in this section will enable children to become 'sustainability heroes.' Every suggested task is aimed at the children themselves. If possible, let a fluent reader read out each activity to the group. If not, explain the activity to them and give them the appropriate support to complete it. In trying out these 'sustainability hero' activities, we have found that children feel very passionate about tackling them and simply need a little sensitive steering to ensure that they don't offend or disrupt adults doing their daily work!

Eco-friendly practice

1 Eco inspection

An eco inspection will tell you about anything that is going on at your school or setting that could have a negative effect on the environment. To become an eco inspector, grab a notepad or a tablet and record all the different processes that happen in a day, thinking about the impact on the environment. You might want to look at:

- How people travel to your school or setting.
- Printing and photocopying.
- How rubbish is sorted.
- Food waste.
- Buying supplies.

You could split into groups to cover different times and different locations. Ask office staff politely if you can see what goes on. When you have all your data, sit down together and decide which areas you want to focus on. For example, you might want to look at waste generated at lunch time, or tree planting. Tell everyone in your school or setting what you found and what you plan to do.

Did you spot any eco heroes? Maybe someone is doing something differently and you need to shout about it!

Eco-friendly practice

2 All that glitters...

Glitter can be popular in schools and other settings, especially at Christmas. However, it tends to end up in our waterways and then in the sea. Glitter makes up 92.4% of the 5.25 trillion pieces of plastic in the ocean. Not only does it take thousands of years to biodegrade, it can be mistaken by plankton for food and plankton is then eaten by larger fish. Scientists have decided that we should have a total ban on glitter and they say that even biodegradable glitter can be harmful.

Think about how you can make sure everyone in your school or setting is aware of the damage that glitter can do. What is the best way to get your message across? Help adults to see the positive difference they can make by not buying glitter.

3 Single-use plastics

Plastic items can take thousands of years to break down so we need to avoid plastics that don't get used more than once. Carry out an investigation and see where and when single-use plastics are being used. Towards the end of the day, put on plastic gloves and empty out non-food waste bins onto a plastic tarpaulin. What single-use plastics are there? How could you eliminate these in the future?

Look in and around your school or setting: plastic straws might be used for crafts or for daily milk. Watch out for small plastic bottles for milk. Milk can be poured from large bottles into cups, reducing the number of bottles to be produced and recycled.

The dinner hall is a good place to track down single-use plastics: replace cling film with foil which can be washed and recycled. Sauce sachets being given out with hot dinners can be replaced by large bottles of sauce. Why not organise a regular 'plastic-free lunch' for children who bring a packed lunch? On that day, encourage everyone to try to make sure their packed lunch does not contain any food that comes in a plastic container or plastic wrapping.

Don't forget about hidden plastics in teabags, chewing gum, baby wipes, stickers and crisp packets. Even some paper cups contain plastic.

Could you set some targets to reduce single-use plastics? Have a look at Activity 24: you could design posters to inform others.

Eco-friendly practice

4 Reduce laminating

When I first became a teacher, I'm ashamed to say that I spent many evenings laminating resources to use in class; lettering for displays and names for pegs in the cloakroom. Some of these laminated resources were used for many years but some simply went into the trash. Unfortunately, it is almost impossible to recycle laminated paper and it may never biodegrade, so those pretty laminated sheets could be with us forever. If we can persuade adults to cut down on laminating, it saves money and also helps the environment.

Adults may feel that laminating paper is a good way to cut down on printing. For example, if you laminate a board game printed on paper, it can be used again and again. However, paper can easily be recycled, and it's best for the environment to avoid producing more plastics, especially as a laminator uses harmful chemicals and heat to seal a document.

Find out how many laminators there are at your school or setting, and how many laminating sheets are bought each year. Is it possible to reduce this number?

Look around your school or setting. Where do you see laminated sheets? Make a list.

Here are some eco-friendly replacement ideas for things that are often laminated:

- Names by pegs in the cloakroom. Instead, wooden 'cookies' could be painted with blackboard paint with the child's name and the year written on the cookie with a blackboard pen. Attach to the wall by the child's peg using Velcro. Children love to take these home and they can even put them on the tree at Christmas.
- Sometimes, important documents must be displayed. A4-sized magnetic document frames could be used instead and these can be used again and again.
- Children's names for self-registering in Early Years. Each child's name could be written on a white stone. The children then put their stones in the basket to show they are present.

Eco-friendly practice

5 Reduce printing

On average, a school prints an estimated one million sheets of paper per year. If we can reduce this number, it's good for the environment and saves your school or setting money too.

Raising awareness of the issue can make a difference. You could make posters using scrap paper urging adults to 'think before you print.' Try to get the message across that, if they can, adults should find alternatives to printing. It's also worth encouraging adults not to print in colour unless it is essential. This cuts down on the amount of chemicals being used, miles travelled and the waste generated from toner, ink cartridges etc.

6 Shred it!

Schools and other settings often need to shred documents and that shredded paper can be put to good use. You can use shredded paper to make papier-mâché or simply put it in the compost bin to help make rich fertiliser. People in the local community may be keen to take it to use in chicken coops or as animal bedding. Find out if any local animal charities would like a donation of shredded paper. For example, organisations which help hedgehogs are generally keen to receive shredded paper.

Eco-friendly practice

7 Paper towels

Does your school or setting use paper towels for children to dry their hands? Even if they are made from recycled paper, they cannot be recycled once they have been used. To make one tonne of paper towels, 17 trees must be cut down and 20,000 gallons of water are used.

Can you find out how many towels are used each day? (You could ask cleaning staff how many packs of towels are put out each day.) You may be shocked by the number. Sometimes, each child will use more than one towel to dry their hands each time.

Find out if paper towels can be removed completely and replaced with hand dryers, which are more eco-friendly.

If that's not possible, make signs encouraging other children not to use too many paper towels each time. Ask the cleaning staff if the number of towels being used each day has reduced. If so, celebrate everyone's efforts!

8 Felt tips and whiteboard pens

What happens to used pens at your school or setting? Do they go straight in the bin? Some pen manufacturers run schemes where you can take old pens to a collection point and they will recycle them for you. These schemes sometimes help charities too. Set up pen recycling boxes and keep track of how many pens you have saved from landfill.

Eco-friendly practice

9 Water butt

Having plants in the outdoor area is better for us, and plants and trees soak up some of the CO2 that we produce. Plants need water but we need to remember that water is a valuable resource – using too much water harms the planet. Invest in a few water butts to collect rainwater so that you are using as little tap water as possible to water plants.

10 Stop the taps running

Clean drinking water is precious. One billion people globally have no access to clean water and two million people die every year from water-related diseases. The water that comes out of our taps has been cleaned using processes that release carbon dioxide into the environment so we need to make sure we don't waste water. Running the taps for longer than necessary is bad for the planet. Make signs to remind other children to turn the taps off when water is not needed.

Eco-friendly practice

11 Switch off lights/plugs

Leaving electrical items on when no one is using them can be dangerous. It wastes money and energy as well as increasing carbon dioxide emissions. Make signs to remind adults that they must turn off all the electrical items possible before going home.

12 Cleaning materials

It's important to keep spaces clean and hygienic but many cleaning products can be harmful to the environment. There are alternatives that are just as good, and sometimes they are even cheaper to buy.

Find out which person in your school or setting is the best person to talk to about how eco-friendly your cleaning materials are and have a chat with them or write them a letter.

13 Meat-free days

Eating less meat is one of the best things you can do to help the environment. Meat production uses much more water than growing plants and it also takes more land to produce meat. A huge amount of harmful greenhouse gases comes from cows and other livestock, who burp and fart methane into the atmosphere.

Why not have a vote? Present the reasons why it would be good to have regular meat-free days. Put out two boxes from the recycling: one for meat-free days and one against. Put out pieces of scrap paper or used bottle top lids and explain clearly how to vote. When everyone has voted, count up the tokens to see how they voted!

To organise meat-free days, you may need to speak to someone at your school or setting who has the power to change the menu. Encourage children who have a packed lunch to bring in a meat-free lunch box on those days.

Eco-friendly practice

14 Composting

Composting food waste can drastically cut down on the waste going to landfill and you don't need a huge amount of room for a compost bin. A compost heap will also attract useful minibeasts.

You can build a simple compost bin using pallets – ask people in your community to donate some. Otherwise, you can buy a plastic compost bin for your outdoor space (sometimes people locally give these away).

Keep a small compost caddy indoors where you put food scraps, and empty this into the compost bin outdoors when it is full. Fruit waste, teabags, ripped up cardboard, shredded paper and fallen leaves will do well in your compost bin. Never put cooked food, meat or dairy products in. Remember to turn your compost regularly using a garden fork so that air gets to it.

A wormery is a brilliant way to get rid of food waste because the worms convert it into fertiliser, as well as putting air into the soil. You can make your own wormery with plastic bottles, compost and sand. Just don't forget to feed the worms because they will eat half their weight every day. Don't keep worms in a homemade wormery for too long: release them into a suitable area where they will help plants and help us.

Copyright material from Sarah Watkins (2023), *99 Eco-Activities for Your Primary School*, Routledge

Eco-friendly practice

15 Travel to school

Walking, cycling, scooting or taking the bus to school cuts down on pollution, congestion and damage to the environment. It also means safer roads. Can you find out how many families regularly travel to school without using a car? It would be good to celebrate their efforts.

You could make videos or signs to encourage families to take part in a regular 'clean-air travel' day, whether this is once a week, once a month or once a year.

Keeping the engine running during drop off puts children and the planet in danger. Think about how you could get the message across to adults to switch off their engines while parked.

16 Out in the community

Don't limit your campaigning – get your message out to the wider community! Maybe focus on a local issue such as litter outside a local supermarket or safer cycling routes. Think about who your audience is and how best to reach them. Involve local media to magnify your message. You might be surprised by how supportive people are and how successful you are in achieving your aims.

Another important activity to carry out in the local community is a litter pick. You can involve parents, carers and grandparents. A set of litter pickers doesn't cost much and people tend to take more pride in a litter-free area.

Eco-friendly practice

17 Celebrate environmental-themed days/weeks/months

Raise awareness of climate change at your school or setting by celebrating days, weeks or months that are linked to the environment. Here are a few examples:

RSPB Birdwatch (January/February)
World Wildlife Day (March)
World Water Day (March)
Earth Hour (March)
Earth Day (April)
National Walking Month (May)
Hedgehog Awareness Week (May)
World Bee Day (May)
World Turtle Day (May)
World Oceans Day (June)
Clean Air Day (June)
Recycle Week (September)
National Tree Week (November/December)
Outdoor Classroom Day (November)

18 Invite beneficial bugs

Even in a small space you can attract pollinators by planting herbs like oregano, rosemary, lavender and chives in pots. If you have a larger space, you can plant buddleia and also create 'no-mow' areas where insects can thrive. You could write to your local gardening centre or supermarket to ask them to donate plants.

Copyright material from Sarah Watkins (2023), *99 Eco-Activities for Your Primary School*, Routledge

Eco-friendly practice

19 Glue sponges

Millions of glue sticks are used each year in UK schools. Many types of glue stick can now be recycled, but unfortunately most end up in landfill. It's better to avoid buying single-use plastic as much as possible. Although you do need a plastic container to make a glue sponge, you will never need to replace it, as long as you follow the simple rules below. No more hunting down glue stick lids!

As well as being simple to make, glue sponges are also much cheaper in the long run and you use less glue. I have found that papers glued in using a glue sponge stayed in books longer than those glued in using a glue stick – this method is stickier!

If a lot of glue sticks are used in your school or setting, could you replace them all with glue sponges? You'd be doing the planet a big favour.

 What you need

- A plastic container with a lid. The container needs to be just large enough to fit the sponge. You could recycle a plastic food container.
- A kitchen sponge. I've tried different types and the type shown in the pictures works best.
- PVA glue.
- Tea tree oil.

Eco-friendly practice

 ### *What you need to do*

1. Carefully pour a little glue onto the bottom of the container until the bottom is covered.
2. Place your sponge on top of the glue.
3. Pour another layer of glue onto the sponge.
4. Add a few drops of tea tree oil to reduce bacteria.
5. Put the lid onto the plastic container.
6. Leave it overnight. (You can use it straight away but it works much better once you have let the glue soak in.)

If you want smaller glue sponges, you can use a smaller plastic container and cut the sponge to fit.

How to use:

Just take the piece of paper you want to glue and gently press onto the glue sponge. For example, if you want to glue a page into a book, press each corner of the paper onto the glue sponge then press the page into the book.

Keeping your glue sponge full and fresh!

If you notice your glue sponge is a little dry, try spraying a little water on it. Every now and then, you may want to add a few drops of tea tree oil to keep it fresh. At least once a year (particularly if the glue sponge will not be used for a few weeks), take the sponge out and wash both the sponge and the container with soap and warm water. Then refill with glue as needed.

Eco-friendly practice

20 Make a terrarium

 ### Why?

Plants are good for our physical and mental health. Terrariums, or mini indoor gardens in a jar, need very little care and are a good example of how to grow plants differently. If the jar is closed, the plant will release water vapour which is then reabsorbed by the plant: you will only need to water it every few months. If it is open, the plants will need to be watered every few weeks.

 ### What you need

- Clean, empty glass jars.
- Succulents, air plants or other plants.
- Some dry moss.
- A few small stones.
- Compost.
- Optional: chopsticks.

 ### What you need to do

1. Place some moss in the bottom of the jar.
2. Put some small stones on top of the moss. (These two layers help with water drainage.)
3. Add the compost.
4. Plant your plants. It's best to use plants that do not need much water. Succulents and cacti are ideal. If you are using succulents, you can separate larger plants into smaller sections. They are very simple to divide – you simply pull a section off and plant it.
5. You might want to add other interesting items such as glass beads.
6. Use a sprayer or a mister to keep it watered. (Be careful not to overwater.)
7. Enjoy your garden in a jar!

Copyright material from Sarah Watkins (2023), *99 Eco-Activities for Your Primary School*, Routledge

Eco-friendly practice

 Extension activities

- Grow cress on the windowsill by sprinkling cress seeds on cotton wool in a recycled container.
- Set up a kitchen scraps indoor garden: carrots, spring onions and celery grow well. Simply put the part you don't eat (the top of the carrot, the bottom of the spring onion and celery) in a small dish with a little water in. When the vegetable starts to sprout, you can plant it in a pot of compost.

2
RECYCLING AND UPCYCLING

This section is all about seeing waste differently. Too often, we don't think too deeply about how we dispose of things. Are you a 'wishful recycler'? Do you put things in the recycling when you are not sure if they actually can be recycled? The 'out of sight, out of mind' strategy will no longer cut it, unfortunately. Our landfill sites have limited space and landfill accounts for around a quarter of the UK's methane emissions. Towns such as Kamikatsu in Japan are already thinking differently. The families who live here aim to recycle 100% of their waste and this was the first area in Japan to pass a zero-waste declaration. They take apart everything destined for the trash and think carefully about how each element can be recycled or upcycled. Let's do trash differently.

Recycling and upcycling

21 Visit a recycling plant

A tour of your local recycling plant provides an unmissable opportunity to find out how waste is managed and see exactly how some waste is recycled. These tours are often free: you just have to organise transport.

22 Join a Scrapstore

You can find Scrapstores all over the UK. They collect items from local businesses that were destined for landfill. Schools and community groups can visit to get brilliant crafting objects and even big items like netting which can be used to build dens.

23 Organise a swap shop

One person's trash is another person's treasure! Why not organise an event where people can pass on things they don't want or need in exchange for something else? This reduces what is going into landfill and encourages recycling habits.

You could ask families who are linked to your school or setting to donate toys and books that are no longer needed then invite them in after school to choose a few items each. You could consider giving people a voucher for every item they donate to make it fairer. They can exchange these vouchers for items to take away.

Recycling and upcycling

24 Make awareness raising posters

🌍 Why?

Posters can be a great way to get your point across. They give people information and advice and usually have a strong message. What did you find out in your eco audit? What issues do you need to tackle?

Copyright material from Sarah Watkins (2023), *99 Eco-Activities for Your Primary School*, Routledge

Recycling and upcycling

What you need

- Paper or card (this could be from the recycling bin).
- Old magazines.
- PVA glue.
- Marker pens.

What you need to do

1. Decide on the message you want to get across.
2. What is the best way to say it? Think of a slogan that will stick in people's minds.
3. Cut or rip different coloured sections out of the old magazines and plan a design that will engage and inform.
4. Put up your posters in places where your target audience will see them.
5. Recycle your posters when they are no longer needed.

Extension activities

Make a huge collaborative eco-hanging using recycled fabrics for a public space. Lots of companies like accountants or law firms with a shopfront are keen for artwork to display.

Recycling and upcycling

25 Junk modelling

🌍 Why?

Junk modelling helps us to see the objects in the recycling bin in a different way: a used drinks carton becomes a house and a toilet roll becomes a fish!

Recycling and upcycling

What you need

- Clean, used packaging such as kitchen towel rolls, cardboard boxes of different sizes, plastic bottles and lids.
- String.
- Masking tape.
- Glue.
- Split pins.
- Scissors.
- Optional: paints and felt-tip pens.

What you need to do

1. Look at all the materials and decide what you want to make. You could choose to decorate your work with paints or felt-tip pens. Once your creation is finished, you could create a junk gallery where all the masterpieces can be displayed.
2. You could have a theme like robots. Ask parents and carers to send in certain items from their recycling like clean foil and bottle top lids.
3. You could also work as a team to make one big piece of artwork. For example, you could make one huge 'town.' You will need to talk about what you might find in the town, such as a park or a pond, and work together to make it.

Extension activities

- Why not look at artwork by artists such as Steve McPherson, Guerra de la Paz, Gilles Cenazandotti and Tess Felix? Their artwork may inspire you to create something spectacular from junk.
- Create a display for a public space such as the library. Contact the right people and ask whether they would like to display some unique eco-artwork for the public to see. Make a small sign for each creation.

Recycling and upcycling

26 Recycled wreaths

 Why?

It's a great feeling to produce something beautiful from materials going in the bin that cost nothing. A wreath makes a perfect decoration for the home or classroom and also makes a great gift.

 What you need

- Paper plates.
- Recycled cardboard (for example, from cereal boxes).
- Paint.

Copyright material from Sarah Watkins (2023), *99 Eco-Activities for Your Primary School*, Routledge

Recycling and upcycling

- Glue gun or PVA glue.
- Scissors.
- Optional: scraps of ribbon.

 What you need to do

1. Cut the centre out of the paper plate.
2. Cut a leaf shape out of cardboard. This will be your template. It should fit nicely onto your plate.
3. Draw around this many times on the cardboard and cut the leaf shapes out.
4. Paint the plain side of each leaf. You can make them as colourful as you like.
5. Leave them to dry.
6. Start to glue the leaves onto your paper plate and keep going until you have a wreath!
7. You could add scraps of ribbon to decorate.

 Extension activity

Cut a circle from a piece of recycled cardboard and cut a hole in the centre. Cover the cardboard circle with PVA glue and stick strips of leftover fabric to this to create a wreath.

Recycling and upcycling

27 Make a recycled kite

 Why?

This is a fun way to use recycling and you can also test your kite flying skills! How high can you fly?

 What you need

- Two sticks or thin branches per kite.
- Masking tape.
- String.
- Scissors.
- Marker pen.
- Used plastic bags. (Otherwise, you could try using thin material or bubble wrap sheets.)
- Other strips of recycled material to attach to the bottom of the kite.

Recycling and upcycling

 What you need to do

1. Make a cross from the two sticks and attach them together with tape.
2. Cut open your plastic bag and spread it out.
3. Put your stick cross on top of the bag.
4. With a marker pen, draw a kite shape around the cross.
5. Cut this out and attach to the cross with tape.
6. Cut a hole out of the bottom of the kite and attach a long length of string.
7. Attach some streamers.
8. Now try flying your kite!

 Extension activity

Try making a kite from a paper bag. Attach four short lengths of string to the corners and tie these four strings to one long string. Attach streamers to the other end.

Recycling and upcycling

28 Recycled marble run

 Why?

Here's a chance to get creative with items from the recycling bin and build your problem-solving skills! This would make a great challenge for teams.

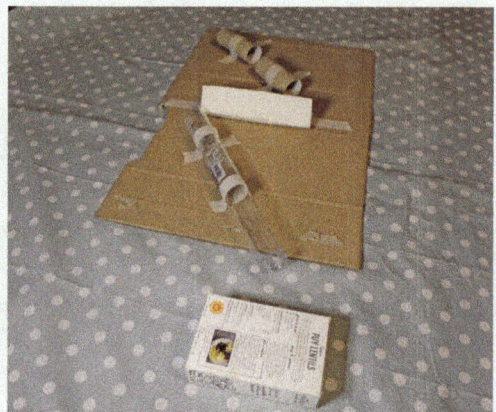

 What you need

- Marbles.
- Masking tape.
- Scissors.
- Clean packaging from the recycling bin. Tubes are a must (toilet rolls and kitchen towel holders, for example). Other packaging that works well: cardboard, cereal boxes, yoghurt pots, cardboard juice boxes, toothpaste boxes.
- If you can get hold of very large cardboard boxes, these are useful for making a vertical marble run.

 What you need to do

1. Decide whether you are going to work in teams or pairs or on your own.
2. Start to plan your marble run. You might want to draw out your plan on paper first.
3. Do you want it to be vertical or horizontal? If you want it to be horizontal, you will need to think about how to combat gravity! You could use tubes to raise the marble run up and gradually reduce the height of the tubes so that the marbles can roll downwards. If you are creating a vertical marble run, you could create it inside a box or on top of a board.
4. Now start building and testing. Cut your pieces to size and tape them in place.

5. Test out your marble run and adjust if the marble gets stuck or slows down too much.
6. Think about how the marbles will be collected at the end. You could consider a small box to catch them.
7. Swap round and try each other's marble runs!

Extension activities

- You could decorate your marble run using paints, pens or crayons.
- Can you put in small bits of card that can change the direction of the marble?
- Try a range of objects such as pom-poms and rubber balls.

Recycling and upcycling

29 Straw shooter

 Why?

These are quick to make and the finished product will make you try to beat your personal best! It's a fun way to use rubbish.

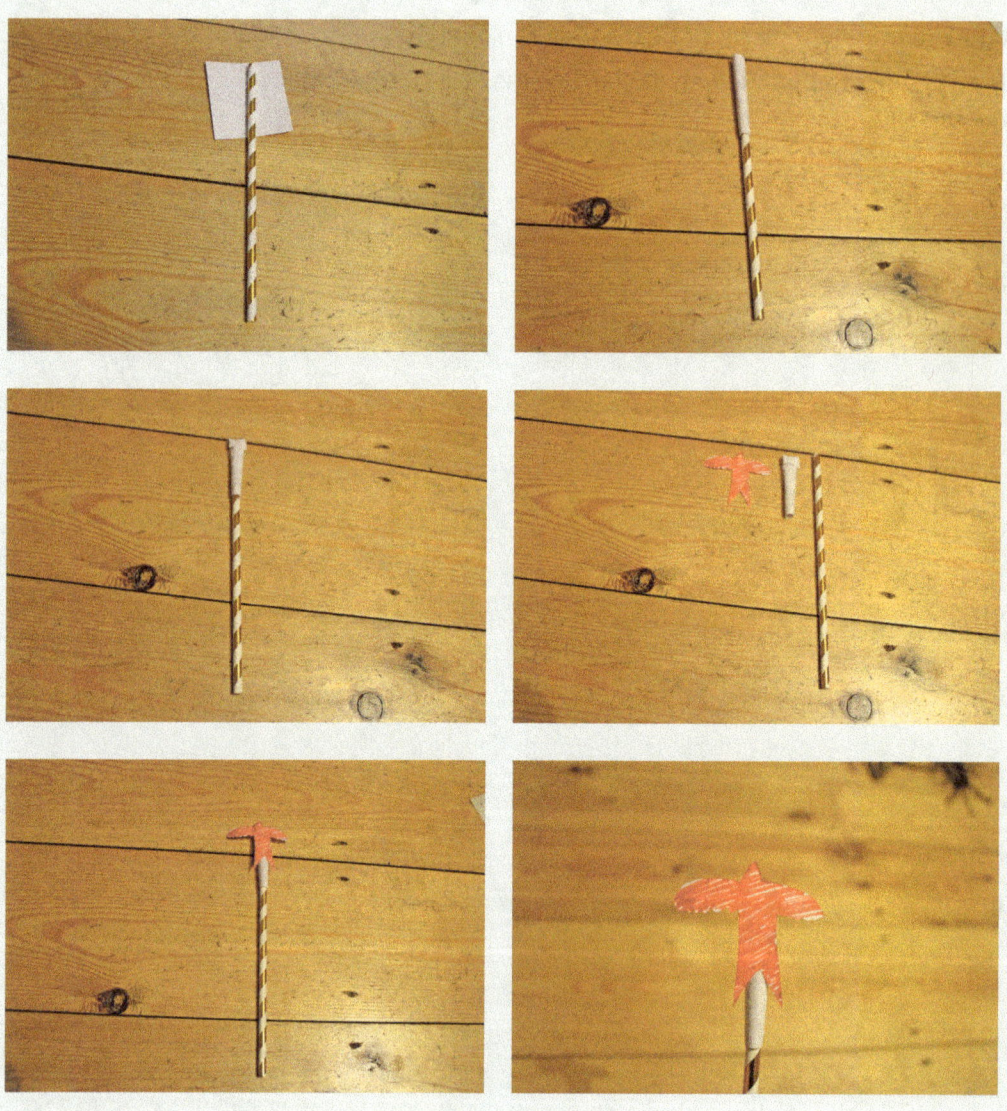

 What you need

- Paper straws.
- Paper from the recycling.
- Pens and pencils.
- Sellotape or masking tape.
- Glue.

What you need to do

1. Cut a piece of paper that is about one-third of the size of the straw.
2. Place the piece of paper on the table and put the straw on top, leaving about one-third of the paper on top.
3. Now roll paper around the straw. It needs to be fairly tight but not too tight!
4. Put some tape around the paper so that you have a small paper tube on your straw.
5. Fold over the top of the paper tube and tape it in place.
6. Take off your paper tube and put it to one side.
7. Take another piece of paper and draw a flying creature on it. This needs to be glued to your paper tube so don't make it too large. You could draw a bee, a fly, a bird, a dragonfly or another creature. Colour this in and make it bright and eye-catching.
8. Stick this to one side of your paper tube.
9. Put your paper tube back onto the end of your straw and blow!
10. How far can you shoot your flying creature?

 ### Extension activities

- Why not set up a race track? Get an old roll of wallpaper or a large piece of cardboard and stand at the end. How far can you get your creature? Mark where it landed and then encourage others to have a go.
- Have a go outside. Is it easier or harder?

Recycling and upcycling

30 Tin can windsock

 Why?

We see the effects of climate change in our weather. This activity will provide you with a beautiful decoration, using recycled items. It would brighten up a garden or a classroom.

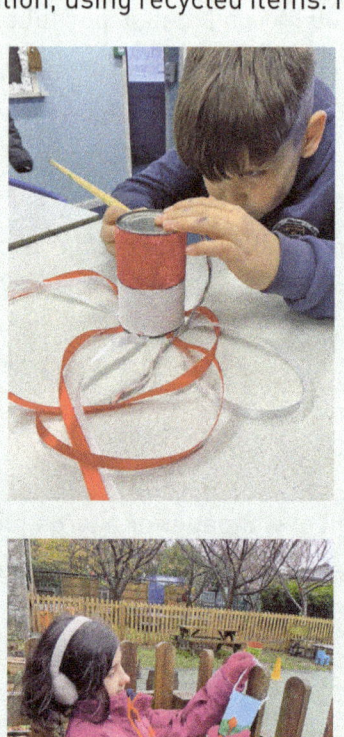

What you need

- Empty tin cans with the labels removed. (Check for sharp edges.)
- Acrylic paint.
- Aprons/old shirts and paintbrushes.
- Newspaper or tablecloths to protect surfaces from paint.
- Leftover ribbons or strips of material. If you have some used plastic bags, you could cut strips from these. You could also use leftover crepe paper.
- Hot glue guns. (Or masking tape or duct tape and scissors.)
- String or twine.

 ## What you need to do

1. Decide on the pattern you want. Paint your tin can and leave it to dry.
2. Cut strips of fabric or ribbon to the right size.
3. Once your tin can is dry, glue (or tape) the ribbons or fabric strips around the inside of the open end of the can so that they hang down.
4. Cut a length of string or twine as a hanger and glue the ends to the top of the tin can.
5. If you want to, you could also glue pieces of fabric to outside of the tin can.
6. Display!

 ## Extension activities

- Cut stencils out of recycled cardboard and tape these to the can with masking tape. Paint over your stencils to create a cool pattern.
- Instead of the strips of fabric or ribbon, use beads tied on wire.

Recycling and upcycling

31 Newspaper fort

 Why?

Making a newspaper fort is a fun activity and it will also test your problem-solving. It's a great way to see how to strengthen newspaper from your recycling bin.

Recycling and upcycling

 What you need

- Newspapers. (You could ask parents and carers to donate them.)
- Masking tape.
- Optional: pieces of fabric/a stapler.

 What you need to do

1. Take two sheets of newspaper and place them on top of each other.
2. Start at one corner and roll to the opposite corner. Roll as tightly as possible so that it is stronger.
3. Put tape on the tube to hold it together.
4. Once you have three newspaper tubes, attach them in a triangle shape. (You might want an adult to staple the triangle.)
5. Now decide how you want to build your fort by attaching triangles to each other.

 Extension activities

- Have a 'camp out' in your fort!
- Make forts out of large, used cardboard boxes.

Recycling and upcycling

 # Make a face

 ### Why?

This is an opportunity to be creative and use your imagination to create artwork from recycled cardboard.

 ### What you need

- Recycled cardboard including old boxes and cardboard tubes.
- Plastic lids.
- Glue.
- Paint.
- Paintbrushes, aprons/old shirts.
- Newspaper or tablecloths to protect tables.
- Optional: wool.

Recycling and upcycling

 What you need to do

Get creative!

1. Decide on the shape you want for your 'face' and cut that out.
2. Cut out other shapes for the features of the face and glue them on.
3. You could use wool for hair.
4. You can choose to paint it or leave it blank.

 Extension activities

- Make your 'face' into a mask.
- Can you make a 3-D 'head'?

Recycling and upcycling

33 Mosaic

 Why?

This activity can bring the community's attention to waste as well as producing a piece of interesting artwork.

 What you need

- Plastic bottle tops of different colours and sizes.
- PVA glue/hot glue gun.
- Canvas or piece of scrap wood.

Recycling and upcycling

 What you need to do

This activity can be really effective if you all work as a team.

1. Arrange your bottle top lids in a pattern before gluing to see how they look. You could arrange smaller lids inside larger lids.
2. Start to glue your lids into place.
3. Once they are in place, let them dry.

 Extension activity

Use old tiles to make a mosaic. (You will need tile adhesive.)

Recycling and upcycling

34 Magnets

 Why?

This is a way to give objects going into the trash a new look. You can create something beautiful from rubbish.

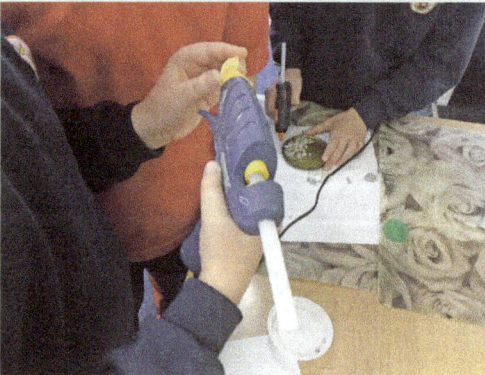

Copyright material from Sarah Watkins (2023), *99 Eco-Activities for Your Primary School*, Routledge

Recycling and upcycling

What you need

- Small magnets.
- Hot glue gun or PVA glue.
- A selection of small loose parts such as small toys, corks, screws, small pieces of wood, scrabble letters, small stones, beads, glass beads, buttons, pom-poms, small shells, small Lego pieces. (You could ask parents or carers for donations.)
- Bottle tops (plastic and metal).

What you need to do

1. Use the lid as a base for your magnet.
2. Glue a magnet to the back.

Recycling and upcycling

3. Arrange small items inside the lid and glue them in place.
4. Enjoy your magnet!

Tracy Forbes from Thomas Coram C of E School: 'I did the activity with our Eco-Councillors and School Councillors and they absolutely loved it! The glue guns were a bit tricky for them, but they managed really well in the end. Lots did the small metal lids and milk lids "treasure" magnets but we'd also got some larger lids and they had fun with these too! They created some small scenes – dinosaur nests and a beach. Then there were the "faces" and a few "natural materials" collections. The children think the book is a brilliant idea and they've asked our librarian to look out for it!'

 Extension activity

Decorate old CDs with permanent markers and then stick a magnet on the back.

Recycling and upcycling

35 Upcycled pencil holder and pencil case

Project 1: Upcycled pencil holder

 Why?

This is a simple and fun way to reuse unwanted containers and it makes a good gift or item to be sold to raise funds.

 What you need

- Some empty tin cans without sharp edges. (You could also use empty, clean jars.)
- Patterned paper (or create your own patterns on plain paper).
- PVA glue.
- Scissors.

Copyright material from Sarah Watkins (2023), *99 Eco-Activities for Your Primary School*, Routledge

What you need to do

1. Lay the tin can side on the paper.
2. Make a mark on the paper and cut a length of paper.
3. Glue the paper on with PVA glue.
4. Put some pens or pencils in your new holder.

Children at Brockwell Junior School:

> 'I thought I could guess how much fabric I needed, I understand now that it's easier to measure what you need first!'

> 'I loved reusing materials to create something new.'

> 'Some people struggled to finish their pots, I liked helping them.'

 ### Extension activities

- Wrap your pencil holder with fabric or wool.
- Make a pencil holder from clean, empty plastic bottles – cut them down with scissors and use a hot glue gun to personalise them!

Project 2: No sew fabric pencil case

 ### What you need

- Some fabric (thick fabric works well).
- A zip.
- Fabric glue or a hot glue gun.

 ### What you need to do

1. Cut the fabric into a rectangle. Lay the fabric out flat. Bring up about three-quarters of the fabric from the bottom and fold it over then bring up the top section and fold it over. Put some fabric glue on the fabric part of the zip and fold over the fabric of the pencil case and attach it.
2. Leave it to dry for a few hours.
3. Turn it inside out and glue the edges together.

36 Recycling old books

 ## Why?

Books in schools are enjoyed by many, many children, and eventually they need to be disposed of if they cannot be repaired. It can be difficult to recycle books because of the glue used in them. Some books, like geography books, become out of date and inaccurate, and so charity books don't want them. Luckily, there are fun projects we can carry out with used books!

 ## What you need

- Old books. For example, books that are out of date or ripped or broken. You could ask charity shops for used children's books, as they often have books donated that can't be sold due to their condition. Old maps are also great for this type of craft – maybe ask families if they have any to donate?
- Glue.
- Scissors.
- Paper and coloured card.
- PVA glue.

 ## What you need to do

Project 1: Greetings cards

1. Think about the design of your card. Who is it for? You might want to make a whole set of cards to sell for charity.
2. You could use cookie cutter shapes to draw around and cut these shapes out to glue onto card. Layering the shapes, one on top of the other, can be effective.
3. If you have a selection of different types of old books, you could experiment with combining different textures. You might want to cut out certain words to use inside or on the front of your card.

Recycling and upcycling

Project 2: Tealight holders

For this project, you will also need clean, empty jars and tealights.

1. Cut up book pages and paste sections onto the glass jar, using PVA glue.
2. You might want to cut a shape out of one piece to show the candle.
3. When the surface is covered, paint over the paper with glue, using a paintbrush.
4. Leave to dry and then put a tealight inside.
5. Never leave a lit tealight unattended. (You may want to use a battery operated tealight.)

Recycling and upcycling

Project 3: Butterflies

For this project, you will need small self-adhesive magnets.

1. Find pages you like and cut out butterfly shapes to be layered up on top of each other.
2. Glue them together in the centre with dots of PVA glue.
3. When finished, stick a small magnet to the back so that you can display your work on the fridge or other magnetic surface.
4. You could try other animals such as bats.

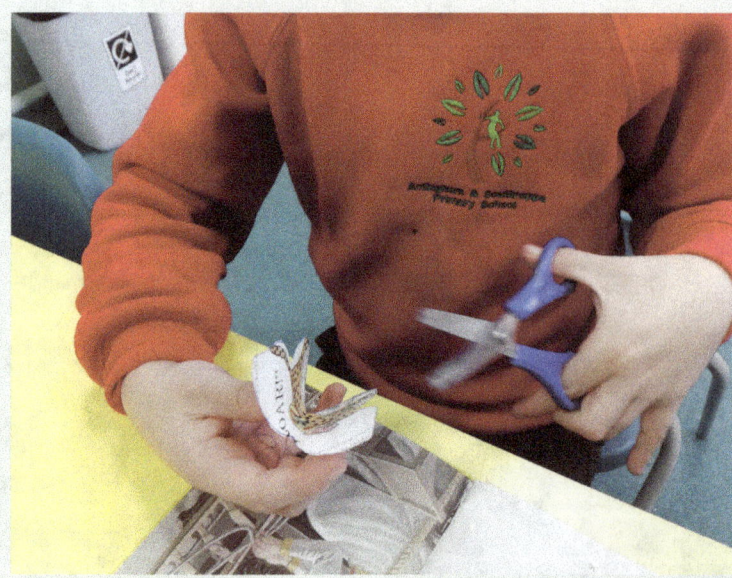

Recycling and upcycling

Project 4: Bunting

For this project you will need string or wool.

1. This makes an eye-catching display. Decide on which shape you want for your bunting and cut these shapes out.
2. To attach the shapes to your string or wool, you could punch holes in the paper and thread the string through or you could use mini pegs or fold the paper over and glue it.

 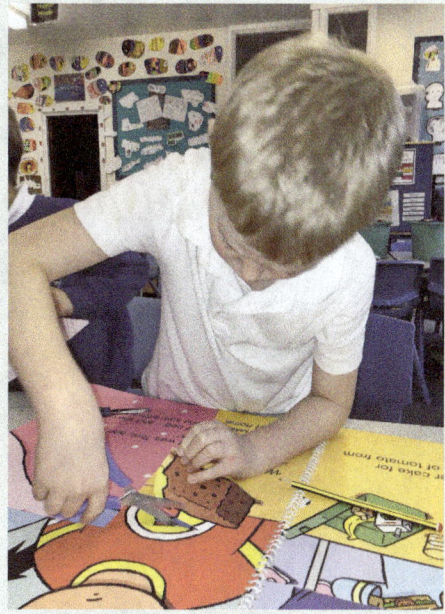

Recycling and upcycling

Project 5: Bookmarks made from books!

For this project, you will also need a hole punch, punch hole reinforcer stickers and card.

1. Find interesting parts of a used book and cut out bookmark shapes.
2. Glue them onto card to make them last longer.
3. You may want to draw a picture on your bookmark.
4. Punch a hole near to the top of the bookmark and stick a reinforcer sticker onto both sides to make it stronger.
5. Tie a short piece of ribbon or string through the hole.

Project 6: Gift tags

For this project, you will also need a hole punch, punch hole reinforcer stickers and card.

1. To make gift tags, cut circles, rectangles or other shapes out of old book pages.
2. Glue them onto card to make them stronger.
3. Cut out pieces of paper for the message and glue these onto your gift tags.
4. Punch a hole in each gift tag and reinforce it with a punch hole reinforcer sticker.
5. Thread a piece of string or ribbon through the hole.
6. Attach to a gift or sell for a good cause.

Recycling and upcycling

37 Recycled notebooks

🌍 Why?

Rather than buying new notebooks, why not reuse old cardboard to create something beautiful and useful. You could consider making a whole batch of recycled notebooks for Nursery or Reception – the children will be delighted!

Recycling and upcycling

 ## What you need

- Cereal boxes.
- Hole punch.
- String or ribbon.
- Paper.
- Paint/felt-tip pens.
- Paper.
- Scissors.

 ## What you need to do

1. Cut a fairly small rectangle out of the cereal box – you need to punch a hole through the centre with the hole punch.
2. Decorate the plain side of your cereal box. You could even use some decoupage for this stage.
3. Cut paper to the same size or slightly smaller than the cereal box cardboard.
4. Put the paper inside the cereal box cardboard and fold the whole thing in half.
5. Punch two holes in the centre. Thread a length of string or ribbon through the holes and tie it together.

 ## Extension activity

Decorate your recycled pad with a homemade stamp. Take a piece of recycled polystyrene and use a pencil to poke a simple pattern into it. (You could 'draw' a leaf or a flower, for example.) Press this onto an ink pad or paint it. Then press it down hard on the cover of your notebook.

Recycling and upcycling

38 Comic decoupage

We don't often reuse comics but they are perfect for decoupage. The only danger is that you may find yourself reading sections of the comic as you cut them up!

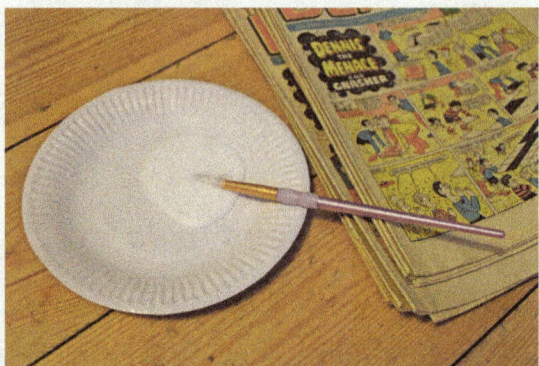

What you need

- Old comics.
- Shoe boxes.
- PVA glue.

What you need to do

1. Cut up sections of the comic and paste them onto the outside of the shoebox until it is covered. Use a paintbrush to add a thin layer of glue on top of the comic scraps.
2. Leave it to dry. Think of what you will store in your new box!

Extension activity

Use recycled magazines to transform old tins that used to hold biscuits or chocolates. Cut or tear small sections and paint these on the outside of the tin with PVA glue.

Recycling and upcycling

39 Recycled Christmas gifts

🌍 Why?

Christmas is a time when people can cause a negative impact on the environment. Thousands of tonnes of food are thrown away, and metres and metres of wrapping paper are bought, used and thrown away. Christmas cards also contribute to waste.

Project 1: Wooden cookies

🛠 What you need

- Wooden slices with a pre-drilled hole. (If you have a hand drill, you could drill a hole, working with an adult.)
- Paint.
- String or ribbon.

Copyright material from Sarah Watkins (2023), *99 Eco-Activities for Your Primary School*, Routledge

What you need to do

1. Choose a design and paint your wooden cookie.
2. Leave it to dry.
3. Once it is dried, tie a short length of string or ribbon to it.
4. Hang up and enjoy!

Project 2: Snowflake decoration

What you need

- Beads.
- Pipe cleaners.
- Ribbon.

What you need to do

1. Cut the pipe cleaners in half.
2. Fold three of the sections around each other to make a star shape.
3. Thread beads onto each section and fold the ends over. Before you fold the last end over, put a small length of ribbon inside the end of the pipe cleaner.

Project 3: Wool Christmas tree decoration

What you need

- Green wool.
- Cardboard.
- PVA glue.
- Mini pom-poms.

Recycling and upcycling

 ## What you need to do

1. Draw a triangle fir tree shape on card and cut out.
2. Cut a long length of green wool.
3. Tie it around the centre of the card tree.
4. Drip glue all over one side of the card.
5. Wrap the wool round and round the tree. Tie it in a knot and decorate the tree with pom-poms.

Project 4: Bowl gift

 ### What you need

- A bowl (a plastic one works well).
- Vaseline.
- Newspaper.
- PVA glue.
- Black and gold paint.
- Bowls.

 ### What you need to do

1. Cover the inside of the bowl with Vaseline.
2. Rip the newspaper into strips.
3. Coat each strip with glue and stick it to the inside of the bowl until it is covered. The more layers you have, the stronger your bowl will be.
4. Leave this to dry for at least 24 hours.
5. Carefully ease the dried structure out of the plastic bowl.
6. Paint it.

Recycling and upcycling

40 Penny spinners

 Why?

Penny spinners are a great example of forces in action and also an opportunity to use old cardboard creatively!

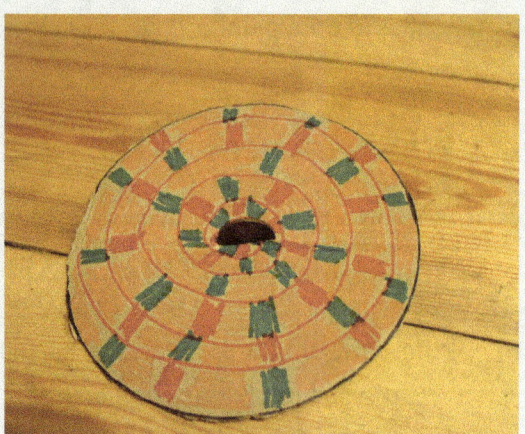

Recycling and upcycling

 ## What you need

- Recycled cardboard (for example, old cereal boxes).
- Scissors.
- Some pennies.
- Felt-tip pens.
- A circle to draw around. (You could use a plastic cup.)

 ## What you need to do

1. Draw a circle on a piece of cardboard and cut it out.
2. Decorate the plain side of the cardboard circle, using felt-tip pens. It's most effective to make circular patterns.
3. Cut a small slit in the centre of the circle for the penny to sit in. You want to be able to push the penny in and for it to stay there without falling out so don't make the hole too big.
4. When your penny is in place, put your penny spinner on the floor or on a table and spin your spinner, watching the colours and patterns change!

 ## Extension activity

Make a paper spinner threaded onto wool or string. Cut two holes in the centre of your spinner and thread the string or wool through. Tie it so that you have a large loop at either side. Pull the two ends tight then relax it and repeat. Watch it spin and hear it hum!

Recycling and upcycling

41 Recycled jigsaw frame

 Why?

Jigsaws are usually thrown away when they have pieces missing and this is a way to put those jigsaw pieces to good use.

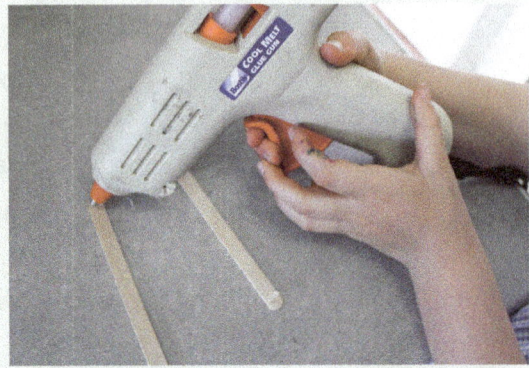
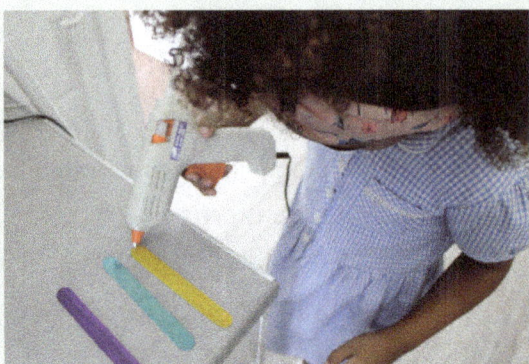

 What you need

- Old jigsaws (you may have some at your school or setting that are missing pieces).
- Lolly sticks.

- PVA glue or a glue gun.
- Felt-tip pens/coloured pencils/paint.

 What you need to do

1. Take four lolly sticks and attach them together using glue so that you have a frame. Leave this to dry.
2. Turn the jigsaw pieces over so that you can see the plain side.
3. Colour these in using your choice of materials.
4. Arrange these on your new frame so that they make a good pattern.
5. Glue these into place.
6. Now you can attach a picture behind your frame using Blu-Tack!

 Extension activity

Use small scraps of fabric and glue them to old picture frames with PVA glue to make a new product. You can pick up small, used picture frames cheaply from charity shops.

Recycling and upcycling

42 Recycled instruments

 Why?

Getting creative with materials from the recycling bin helps us to think about the trash we generate but also helps us use our imagination and achieve challenges.

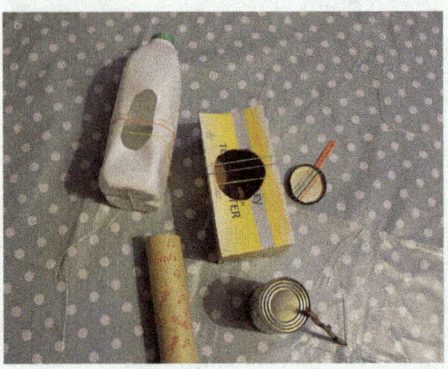

 What you need

- A range of clean, recycled materials.
- Elastic bands.
- Used balloons.
- Dried beans.
- Glue.
- Scissors.
- Masking tape.

 What you need to do

1. Think about which instruments you can make with the materials you have.
2. Here are some ideas:
    - Drums – you could use a used balloon stretched over a cardboard tube with an elastic band.
    - Shakers made with tins, elastic bands, paper and dried beans.
    - Tambourines made from plastic plates.
    - Rain sticks made from tubes that contained crisps plus rice or dried beans.
    - Guitars made with pots/cardboard boxes and elastic bands.
    - Make trumpets from plastic milk bottles.

 Extension activity

Make a set of shakers for a local nursery. Decorate an old cardboard box for the shakers to be delivered in. This will be a great surprise for the children!

Recycling and upcycling

43 Egg box mobiles

 Why?

We're used to seeing egg boxes from one perspective but cut them open and you open another world of possibilities!

 What you need

- Egg boxes.
- Paint.
- String or wire.
- Sticks or old clothes hangers.

Recycling and upcycling

 ## What you need to do

1. Cut out the parts of the egg boxes that hold the eggs.
2. Paint these different colours.
3. Turn them upside down and cut a small hole in the top.
4. Cut a length of string and thread this through the egg box parts, tying a knot to keep each one in place.
5. Repeat until you have two or three strings.
6. Tie these onto a stick or a clothes hanger.

Mr Shabbir of The Village Prep School, London:

'At first, the girls were very confused to find out that a mobile was in fact not always a phone! But once this was cleared up they were very excited to get started. The girls were not satisfied with only one paint colour for their mobiles and proceeded to spend plenty of time delicately colouring the sides of their mobiles in all sorts of different colours. By the end, all the paint was mixed and the girls used a brand new colour that they had created to finish off their mobiles. After this, we had to let the mobiles dry before continuing our project the next week.

The following week the girls persisted to get their string through the mobile. It took a lot of tries but the girls' sheer determination got them through it. Then together we discovered how to tie knots onto our sticks. The girls saw their creations complete and showed them off with pride! The girls enjoyed shaking their mobiles around at the end but for all the best part was definitely the painting!'

 ## Extension activity

Make a plastic bottle mobile. Cut sections out of different coloured plastic bottles so that you have a good range of plastic rings. Punch a hole in each one with a hole punch and tie them to a stick to make a mobile.

Recycling and upcycling

44 Marine life mobiles

🌍 Why?

The plants and animals in our oceans are increasingly negatively affected by climate change. Using recycled materials to create a marine life mobile can get people talking about what is happening and what we can do about it.

Copyright material from Sarah Watkins (2023), *99 Eco-Activities for Your Primary School*, Routledge

Recycling and upcycling

 ## What you need

- Paper plates.
- Clean, used plastic milk bottles.
- A hole punch.
- Felt-tip pens/crayons/coloured pencils.
- Wool.
- Scissors.
- Permanent marker pens.
- Optional: bottle top lids.

 ## What you need to do

1. Cut a circle out of the centre of the plate.
2. Punch a series of holes out of the edge of the plate, some high and some low.
3. Decorate your plate.
4. Cut three different lengths of wool.

5. Tie the lengths of wool to three or four of the holes and weave these back and forth, across the plate and through the holes. (Leave three holes at the base of the paper plate free of wool when weaving.)
6. Cut large sections out of the plastic bottles. Then cut marine animals such as fish, jellyfish and starfish out of these large sections. Decorate these with permanent markers. Punch a hole in each one and tie each one onto the end of a piece of wool.
7. Tie these to the three holes at the bottom of your dream catcher.
8. Optional: carefully make a hole in the bottle top lids and thread these onto the wool, tying holes to keep them in place.

 Extension activity

Make a sea turtle mobile. Cut out the egg holder sections of an egg box and paint these green. From green card, cut out the shape of a turtle head, body, legs and tail. Make a hole in the top of each egg box 'shell' and attach a piece of string to each one. Glue the shell on the top of the body. Draw eyes on. Attach every piece of string to a paper plate to make your mobile.

Recycling and upcycling

45 Plastic jar lanterns

🌍 *Why?*

Why not make an attractive keepsake from recycled materials?

Recycling and upcycling

 What you need

- Clean, used plastic food containers such as small milk bottles.
- Tissue paper.
- PVA glue.
- Battery operated tealights.

 What you need to do

1. Cut the top off the plastic container, being careful not to leave any jagged edges.
2. Cut up squares of tissue paper and paste these all over the container until it is covered.
3. Leave it to dry.
4. Put in a battery-operated tealight.

 Extension activity

Make milk bottle pen/pencil sorters. Cut a section out of the bottle so that you can slot different coloured pens in.

Recycling and upcycling

46 Recycled T-shirt dog toy

🌍 *Why?*

This is a fun way to re-use old t-shirts and you can either sell these to raise money for your school or setting or donate them to a local dog shelter.

Recycling and upcycling

 ## What you need

- Old t-shirts. (Charity shops often have lots of old t-shirts that they can't sell – ask if they could donate for a good cause.)
- Sharp scissors.
- A ruler.

 ## What you need to do

1. Cut off the sleeves and the head hole.
2. Cut the remaining part of the t-shirt into two parts then cut the fabric into strips that are around 30cm long.
3. Roll each strip into a long sausage.
4. Tie together three of the long sausages at one end. Then plait them and tie them together at the other end.

Top tip from Little Sutton Primary School: It helps when you are plaiting the material to have some tension. Some pupils did this by having a friend hold the end for them or by putting one end under the foot.

 ## Extension activity

Make t-shirt bracelets by cutting strips from old t-shirts and plaiting them together. When you have the right size, tie the ends together.

Recycling and upcycling

47 A recycled socktopus!

🌍 Why?

This is an interesting way to use up socks that have lost their partners and you can sell these to raise money for your school or setting. You could also use unwanted fabrics to stuff your socktopus.

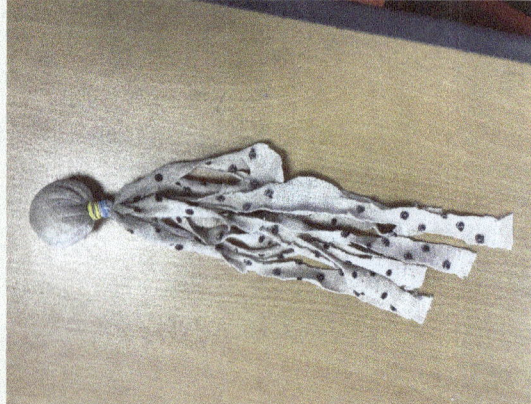

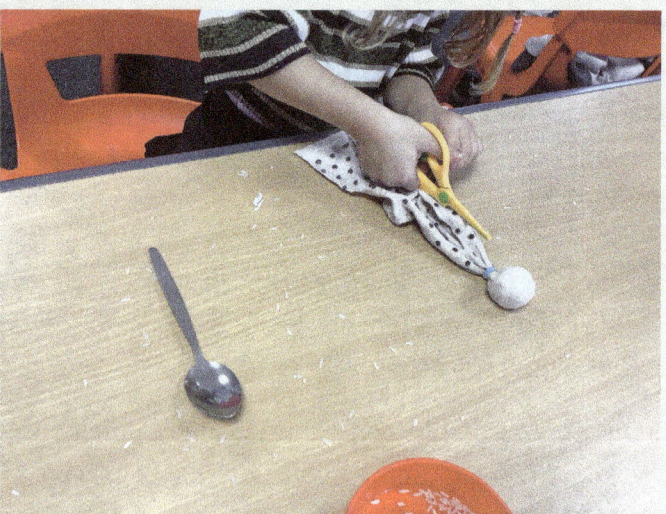

🪴 What you need

- Socks. (Longer socks work best.)
- String or ribbon.
- Googly eyes.
- Something to stuff your socktopus with: you could use cotton wool, scraps of fabric, another sock, used plastic bags.
- Scissors.
- PVA glue.

Recycling and upcycling

 What you need to do

1. Stuff the toe of the sock with your chosen stuffing.
2. Tie string or ribbon round this to ensure the stuffing doesn't fall out.
3. Now take your scissors and cut the sock under the 'head' so that it is in strips or 'tentacles.'
4. Glue your googly eyes onto the head.
5. Enjoy your socktopus!

 Extension activity

Make lavender sock bags by cutting the toes off old socks and filling them with lavender. Leave a little room at the top and tie this with ribbon.

Recycling and upcycling

48 Make recycled crayons

 Why?

When crayons are well-used, the small sections you have left are often thrown in the trash. Why not have a tub where old crayons are 'saved,' ready to be recycled?

 What you need

- Old crayons. They must be wax crayons, otherwise they won't melt properly.
- Knives suitable for children to use.
- Chopping boards.
- Silicone trays for the oven. (Ones with large sections work well. You can choose different shapes.)
- Access to an oven.

This activity needs close adult supervision.

 What you need to do

1. If there is any paper on the crayons, peel this off.
2. Cut the crayons up into very small pieces.
3. Put the broken pieces into the silicone trays. (Only fill them about a quarter full.)
4. Make sure you put in lots of different colours. You might want to group some colours together.
5. Put the silicone trays into the oven and heat until melted.
6. Take out of the oven and leave to cool. (You can put them in the freezer to speed the process up.)
7. When the new crayons have set, pop them out.
8. Enjoy your new crayons!

'I'm going to put all my favourite colours in.'

'Look, the colour is changing!'

Recycling and upcycling

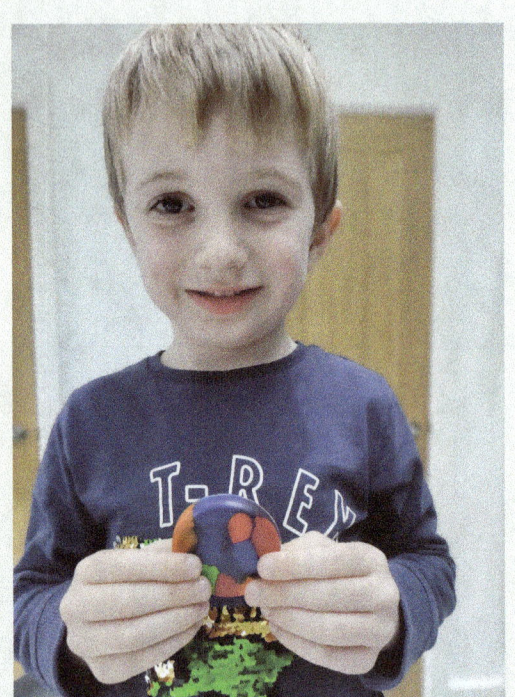 

'It was super fun! When I used it, I drew multicoloured grass and a bus.'

 Extension activity

You could make a batch of these and put them in paper bags to sell at a fundraising event. Or use old crayons to make leaf or bark rubbings.

Recycling and upcycling

49 Make candles

 Why?

This is a way to save old candles and old crayons from the trash, and create a beautiful product.

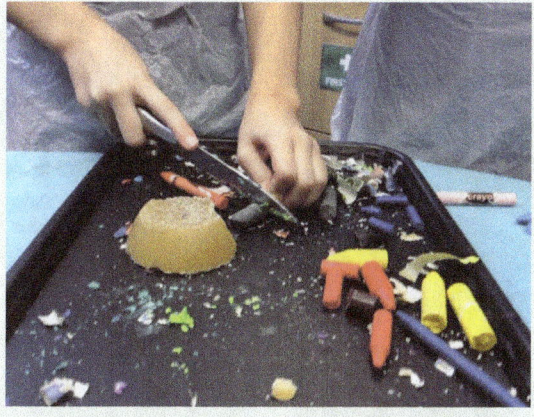

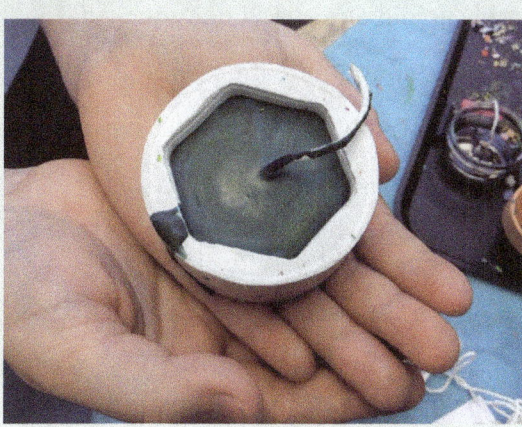

Copyright material from Sarah Watkins (2023), *99 Eco-Activities for Your Primary School*, Routledge

Recycling and upcycling

 ## What you need

The small group activity below is more suitable for KS2.

- Old candles. These could be tealights or long, thin candles. Thicker candles will be harder to chop up but an adult could help.
- Used crayons.
- Long wicks.
- Glass containers. Small, clean, used jars are fine. (Alternatively, you could use silicone cup cases or ice cube trays to make smaller candles.)
- Bowls suitable for going in the microwave.
- A microwave.
- Knives and chopping boards.

 ## What you need to do

1. This activity works well with children in pairs. Adult supervision is required.
2. Cut up the old candles and crayons into very small pieces.
3. Melt a small amount of the old candles in the microwave. One child needs to hold the wick in the jar while another child carefully pours the melted wax into the jar so that the wick is secured. Leave the end of the wick to be lit hanging out of the jar. This can be trimmed later.
4. Mix half candle wax and half crayons together and melt this in the microwave. Pour this into your jar. You could try alternating different colours but this requires you to wait for each layer to dry first.
5. Leave your candle in a cool place to harden. You could decorate the outside.

Stretton Sugwas C of E Academy:

'Wax melts far better in a ceramic bowl. Preferably an old one as it'll get wrecked.'

'Wicks may need to be taped to bottom of the jars to stop them floating.'

'Thicker crayons need to be cut with a sharp knife under adult supervision. Thin ones can just be snapped.'

'Wax honeycombs can be melted. The wax rises to the top, which can then be scraped off, remelted and refined again.'

'Small pudding jars, mini jam jars for breakfast and small terracotta pots work well. We also used moulds which worked really well.'

 ## Extension activity

Roll a beeswax sheet around a candle wick to make a candle. You can even cut up the sheets of beeswax to make multicoloured candles.

50 Make loose parts kits

🌍 Why?

Adults often have small items that are no longer useful and this activity puts them to good use. A loose parts kit can provide hours of creative play for young children and it's a great alternative to a new toy. Never leave very young children unattended with small loose parts.

🪏 What you need

- Loose parts are materials that can be used in multiple ways and combined with each other. Ask parents and carers to donate small, safe items such as unwanted nuts, bolts and washers, curtain rings, small mosaic tiles, beads, old bottle top lids, buttons, old cotton reels, old keys, clothes pegs. Add in tiny offcuts of wood, twigs, shells, lollysticks, pom-poms, elastic bands.
- Small boxes. You could ask for donations of clean, used Tupperware boxes or use small cardboard boxes from recycling.

🌱 What you need to do

1. Put together a selection of small loose parts items in each box which young children can use creatively. You could follow a theme. For example, you could make a 'robot tinkering kit.'
2. Decorate the box to make the kits even more attractive.

These loose parts kits sell well at fundraising events.

💧 Extension activity

Have a 'pick and mix' event where children can come and choose the items they want and take them away in a box.

51 Recycled planters

🌍 Why?

Climate change is already making us think about different ways to grow plants and we are finding that plants are versatile and can grow in many different environments. This is a great way to use recycled materials instead of buying plastic pots.

Project 1: Juice carton planters

What you need

- Empty juice cartons.
- Compost.
- Seeds.
- Scissors

What you need to do

1. Cut one of the long sides off the carton.
2. Poke some holes in the opposite long side.
3. Fill with compost and plant your seeds.
4. Water!

Project 2: Recycled plastic bottle planters

What you need

- A wide range of clean, empty plastic containers.
- Googly eyes.
- Hot glue gun/PVA glue.
- Acrylic paints.
- Scissors.
- Compost.
- Small stones.
- Plants.

What you need to do

1. Look at your container and decide how you want it to look and where you want the plant to come out.
2. Cut it to the correct shape and paint.
3. Glue the parts where you need them.
4. Cut holes for drainage at the bottom and add small stones then compost.
5. Plant your plant and water!

Extension activities

Why not try harvesting your own seeds? Once a sunflower has stopped flowering, you can run your hand over the sunflower head and save the seeds. Make sure they are completely dry and then you can plant them straight away or save them in a cool, dry place until next season. Remember to label the container so that you remember what they are! You could also save seeds from apples and other fruit.

Recycling and upcycling

52 Tin can bulbs

 Why?

Having plants around us is good for our health and wellbeing, and winter flowering bulbs bring a bit of colour while we wait for the spring.

 What you need

- Clean, empty tin cans with the labels removed.
- Compost.
- Acrylic paint.
- Paintbrushes.
- Flower bulbs such as hyacinths or daffodils.
- Small stones.

Recycling and upcycling

 What you need to do

1. Paint the tins and leave them to dry.
2. Put small stones in the base then put compost on top.
3. Plant the bulb in the compost.
4. Water the bulb immediately and then put the tin in cold conditions for ten days.
5. Put the tin on a sunny, warm window sill.
6. Even in winter, your bulbs should flower!

 Extension activity

Garlic bulbs are easy to grow and you can plant them between November and April. Simply split a garlic bulb into its cloves and plant each clove in a pot of compost, with the pointy end facing upwards. Put the pot in a warm sunny spot and water regularly.

53 Newspaper pots

 Why?

Growing our own food is a good way to cut down on food miles, and these pots put old newspapers to use and avoid the need for plastic pots.

Recycling and upcycling

What you need

- Newspaper.
- A cardboard toilet roll from recycling.
- Compost.
- Vegetable seeds.

What you need to do

1. Start with one sheet of newspaper.
2. Fold it in half lengthways.
3. Put your toilet roll on the newspaper. Make sure it protrudes over the top of the newspaper a little.
4. Roll the newspaper along with the toilet roll.
5. Tuck the newspaper firmly into the bottom of the toilet roll to make it secure and stable.
6. Pull the toilet roll out and you can now put compost and seeds in! You could try growing tomatoes, cucumbers or beans.

Vahri Ingleston at Carrongrange High School:

'We felt like the instructions were clear enough to understand and complete without a lot of adult intervention. The pupils enjoyed taking part.

We (teachers) felt that the tasks being built to allow pupils to run them is important, especially when our school focuses a lot on independence.'

Extension activity

You can simply use toilet rolls and plant straight into these (putting the toilet rolls into a tray before you add compost). Egg boxes also make great recycled planters.

Recycling and upcycling

54 Recycled garden decorations

🌍 *Why?*

Making recycled garden ornaments can encourage people to spend more time in the garden, as well as putting recycled materials to use.

Plastic bottle ladybirds

 What you need

- Clean, used plastic bottles with lids.
- Acrylic paint: red, black and white.
- A hot glue gun/PVA glue.
- Black pipe cleaners.
- Scissors.

 What you need to do

1. Cut the bottom off the plastic bottle.
2. Paint it red and let it dry.
3. Paint the bottle top black and let it dry.
4. Paint black dots onto the red bottle section and let this dry.
5. Paint eyes and a smile onto the lid.
6. Cut the black pipe cleaner into two small pieces and glue this onto the lid as antennae.
7. Glue the 'face' onto the body.
8. You could add wings, made of the sides of the bottle and painted red and black.

 Extension activity

Use long strips of material to weave on fences. This creates beautiful recycled artwork.

Recycling and upcycling

55 Make plant labels

 Why?

Planting is an important part of being more eco-conscious. Labelling plants can prevent confusion!

Project 1: Lolly stick plant labels

 ### What you need

- Lolly sticks.
- Felt-tip pens.
- Craft varnish.

 ### What you need to do

1. Write the name and/or draw a picture of the fruit or vegetable on the lolly stick.
2. Paint the lolly stick with craft varnish.

Project 2: Lid plant labels

 ### What you need

- Lids from glass jars.
- Paper.
- Hot glue gun.
- Sticks/lolly sticks/short lengths of cane.

 ### What you need to do

1. Cut a circle from the paper that will fit inside the lid.
2. Decorate the paper and write the name and/or draw a picture of the plant.
3. Glue the paper inside the lid.
4. Using the glue gun, attach the lid to the stick.

Project 3: Blackboard paint plant label

 ### What you need

- Small pieces of balsa wood and lolly sticks.
- Blackboard paint.
- Chalk paint pens.
- Glue gun.

 ### What you need to do

1. Paint the balsa wood with blackboard paint and leave it to dry.
2. Use the chalk pen to write the name and/or draw a picture of the plant.
3. Use the glue gun to attach the wood to the lolly stick.

3
CONNECT WITH THE NATURAL WORLD

This section is all about celebrating the magnificence of our natural world. Connecting with nature helps us value what we already have, and noticing the poppy growing in a crack in the pavement or a robin perching in a tree is good for our mental health. Being in nature, recording nature and protecting nature is great for the environment and helps us too. Even in an urban environment, you can interact with nature and encourage plants and animals. Enjoy different types of weather and feel the benefits of experiencing all the seasons.

Connect with the natural world

56 Make bath bombs

🌍 Why?

A homemade bath bomb with fresh herbs makes the perfect gift...or you can have your own fizzing bath experience!

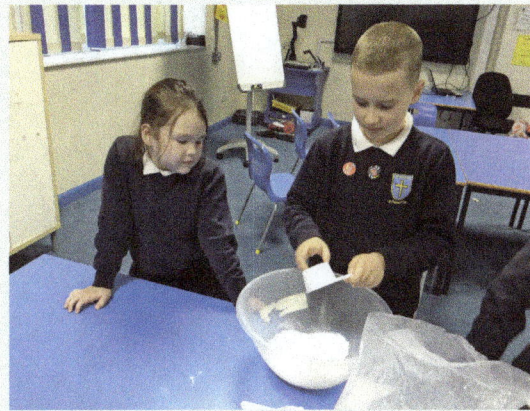

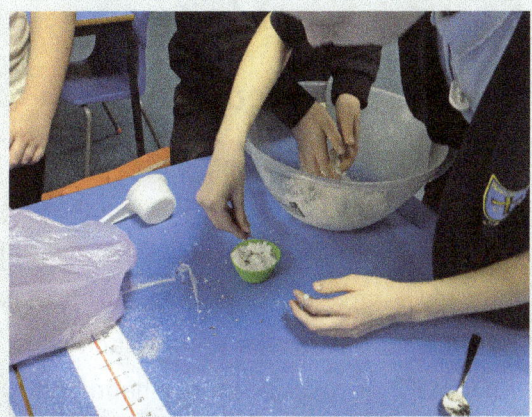

Copyright material from Sarah Watkins (2023), *99 Eco-Activities for Your Primary School*, Routledge

What you need

- Herbs such as lavender, mint, camomile. These need to be chopped up very small. Try growing and harvesting your own.
- Half a cup of Epsom salts.
- One cup of baking soda.
- Half a cup of citric acid.
- Half a teaspoon of water.
- Two teaspoons of sunflower oil.
- Spray oil.
- Mixing bowls.
- Spoons.
- Silicone muffin trays.

Optional:

Natural essential oils.

Natural food colouring.

Fabric scraps and ribbon.

 ### What you need to do

1. Put the baking soda, Epsom salts, citric acid and herbs in a bowl and mix together.
2. In a different bowl, mix the water and sunflower oil together. If you are using them, add a few drops of your essential oil and food colouring.
3. Slowly pour your wet ingredients into your dry ingredients. If you need more water, add a little more. Take care not to activate the citric acid at this point. If you can squeeze some of the mixture in your hand and it sticks together, it is ready.
4. Spray a little cooking spray into your muffin tray.
5. Take a small handful and push it into your muffin tray. Keep doing this until all the mixture is used up.
6. Put the muffin tray into the fridge for a few hours until the bath bombs harden.
7. Pop the bath bombs out of the tray and wrap in spare bits of fabric and tie with a ribbon or wrap them in baking paper.

 ### Extension activities

Add a little lemon or orange peel chopped up very small.

Connect with the natural world

57 Make a bathyscope: an underwater viewer

Only use this in shallow water and under adult supervision.

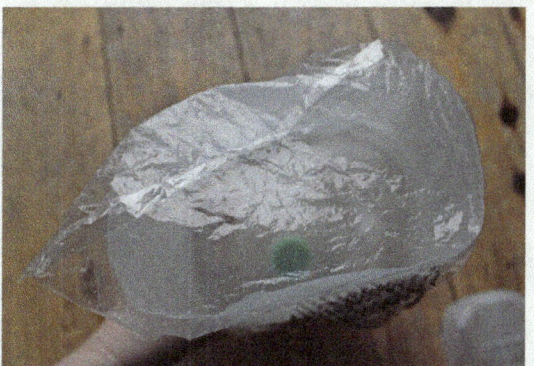

🌍 Why?

This is a great way to check out what is in our ponds and streams – we can see how healthy they are and look more closely at plants and animals.

🛠 What you need

- A clean, used container. A large coffee tin works very well. (You could ask your local coffee shop.) Or a plastic milk bottle (a two-pint bottle is good), a one- or two-litre plastic drinks bottle, or a large yoghurt pot. A large cardboard drinks bottle (this will be less durable).
- Cling film or used clear food packaging.
- Elastic bands.
- Scissors.

Connect with the natural world

What you need to do

1. Remove the bottom of the container. Be careful of sharp edges.
2. Attach a piece of cling film to the base with an elastic band.
3. If you have access to a pond, put the cling film wrapped end of the container into the water and look through the top. The cling film acts as a lens, magnifying your view. This will give you a great perspective of tadpoles and other creatures!

Connect with the natural world

58 Pond dipping

Take care around water. This activity requires adult supervision.

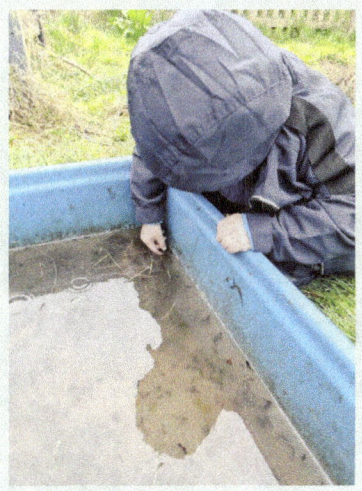

🌍 Why?

As well as exploring the wildlife on land, it is interesting to look more closely at creatures and plants that live in water. You will find most activity in a pond in spring/summer.

🪏 What you need

- Nets.
- Trays for observation.
- Magnifying glasses.
- Collecting pots.

🌱 What you need to do

1. With adult support, dip a tray into the water so that you can observe some of the pond creatures up close. You can also dip the small pots in the water.
2. Dip your net into the water and sweep it in a gentle figure of eight. Carefully turn the net into the tray and see what you have caught.
3. You can use a guide to help identify the creatures you find. Remember to put all creatures back in the pond when you finish.

💧 Extension activity

Keep tadpoles in a fish tank for observation. Put pond water and a few pond plants in the tank. Don't keep tadpoles in a tank for too long and never put them in full sunlight.

Copyright material from Sarah Watkins (2023), *99 Eco-Activities for Your Primary School*, Routledge

Connect with the natural world

59 Make a rain gauge

 ### Why?

Climate change is causing droughts in some areas and floods in others. We need to see rainwater as a valuable resource.

 ### What you need

- Plastic bottles.
- Rulers.
- Permanent markers.
- Scissors.
- Optional: plasticine, small rocks.

 ### What you need to do

1. Cut the top section off your plastic bottle.
2. Place some small rocks in the bottom and then put a little water in so that you have a starting point. Make a little mark at the top of the water so you know where to start measuring. Or you can put plasticine in the bottom instead of the stones.
3. Turn the top of the bottle upside down and put it back into the bottle to act as a funnel.
4. Put your rainfall gauge outside on a flat surface.
5. When it next rains, measure the water that fills the gauge above your mark. (Measure at the same time every day and remember to tip out the water every time you have measured it.)
6. Make a rainfall diary. You can also share your records with the Met Office.

 ### Extension activities

- Measure other weather: measure the temperature at different times of the day. Note down the clouds you can see at different times.
- Create a pine cone weather forecasting station. Place pine cones on a window sill. Pine cones open when it is dry and close when it is raining. Look at the pine cone each day and record whether it is open or closed. Note down whether it is raining.

Connect with the natural world

60 Make a (no glue) twig boat

Why?

This is a great STEM activity using natural materials which will get you thinking scientifically as well as ecologically!

What you need

- String or wool. (For younger children, masking tape might be easier.)
- Scissors.
- Twigs (about five or six per child).
- Leaves.
- Natural decorations such as flowers and moss (optional).
- Some water. The larger the container, the better.

What you need to do

1. Firstly, you need to decide on your mission! Mission ideas include:
    - Make a boat that floats. (Time it for 30 seconds.)
    - Make a boat that survives 'waves' without sinking. (Swirl a spoon in the water to create waves.)
2. When you have decided on your boat building mission, make sure everyone has enough materials and start building! You could work in pairs or on your own.
3. Think about the best construction methods to make sure your boat floats. Weaving the string can work well to stop gaps. Could you plug gaps with natural materials? Do all of the twigs need to go in the same direction? What could make the boat sink?
4. Snap twigs if they are too long, being careful not to snap them too close to anyone's face.
5. When you have a water worthy vessel, finish it off with a leaf sail. You could even decorate the sail to make it yours.
6. Now to test the boats! Think about your mission and see whether each boat passes the test. If it doesn't work, how could you adapt your boat?

Jenny from Copmanthorpe Primary School: 'The children loved this activity. They decided that they needed to go on a stick hunt first and then suggested that they needed to leave them to dry as they had already decided that they wanted to use duct tape. I gave them a variety of ways to connect the sticks and they decided the duct tape would work best. They had some great discussions about how they would connect the sticks. They snapped the sticks, cut the tape and connected the sticks.'

Copyright material from Sarah Watkins (2023), *99 Eco-Activities for Your Primary School*, Routledge

Connect with the natural world

 Extension activities

- See which raft can carry the heaviest 'passenger' without sinking.
- Make a range of boats from recycled materials.

Connect with the natural world

Connect with the natural world

61 Make a water filter

 Why?

Around the world, 771 million people don't have access to clean water close to their home and even more people are losing access to clean water because of changing weather patterns caused by global warming. For example, earthquakes or floods often damage or destroy water supplies. There will be more demand for water as the temperatures continue to rise but this hotter weather can also lead to more harmful organisms in drinking water.

We are lucky to get instant clean water from the tap for drinking and washing.

Connect with the natural world

 ## What you need

- An empty plastic drinks bottle with a lid.
- Cotton wool.
- Stones.
- Sand.
- Materials to make the water dirty such as dirt, leaves and small sticks.
- Two cups: one to stand the upside-down bottle in and one to collect dirty water in.

 ## What you need to do

1. Half fill one of the cups with water then add soil, leaves and other items to make the water dirty. Put this to one side.
2. Put the lid on the bottle and pierce a hole in the lid. Next, carefully cut the bottom off the bottle.
3. Put the section of the bottle with the lid upside down and put in two handfuls of cotton wool.
4. Put a handful of stones on top.
5. Pour two handfuls of sand onto the stones.
6. Then repeat the whole process again!
7. Put the upside-down bottle into a cup and pour the dirty water into the bottle. Watch the filtering process as the water moves through the different layers and gets cleaned. How clean does your filtered water look?

 ## Extension activities

- Try different ways to filter the water: coffee filters, charcoal or gravel.
- Time the process and see how long it takes.

Copyright material from Sarah Watkins (2023), *99 Eco-Activities for Your Primary School*, Routledge

Connect with the natural world

62 Make a nature museum

🌍 *Why?*

Making a nature museum makes us look more closely at nature and think about how to categorise and display different objects.

Connect with the natural world

 ## What you need

- To make a 'takeaway' nature museum, you will need some shoeboxes or similar sized boxes. If you prefer, the container could be tiny to display tiny items!
- For a larger nature museum, you will need somewhere to display the items – maybe a white sheet, a table or a tarpaulin.
- Nature items such as stones, leaves, shells, pine cones. Check that shells do not have live creatures inside.

For older children: a hot glue gun and index cards.

 ## What you need to do

1. Collect a few natural objects that look interesting to you.
2. Display these and talk about why you chose each one. Others might want to ask you questions.

 ## Extension activity

Older children could glue their items into place in their takeaway museum container and research each item. They could write some notes on each one and place the card in their portable museum.

Zoe Sills of Earthtime: 'We used egg boxes for collecting items for our "museum" or display. We call them "treasures!" We do this regularly with children of all ages and find it a really good way of collecting natural items. We sometimes paint or colour in the base of each segment and it becomes a colour-matching scavenger hunt, or for older children who can read we put a list of items they need to collect in the lid of the box for them to go off and find. we often use a white sheet or piece of fabric to create transient art or display things we have collected – they show up so much better than on the ground and

Connect with the natural world

the colours are more distinctive. One of the children suggested we use lolly sticks to label them for our display, "like labelling plants."'

Quotes from children at Earthtime:

'That's a wood pigeon feather, we see lots of those.'

'That oak leaf is MASSIVE!!' (We then talked about the difference between the red oak and our native oak.)

'Why have all the sycamore leaves got spots?' (We talked about the common fungus.)

'The ivy leaves are so shiny!'

'I found some lichen.'

'We found so many conkers! We don't need to put them all in the display but we should put some of the shell too so we can see they go together.'

'These are helicopters!' 'They're sycamore seeds.'

'All the pine cones are different shapes and sizes.' 'We didn't find any of the soft furry ones today.'

63 Nature journaling

 Why?

Outside, you will all notice different things and you can choose to record your thoughts in different ways. A nature journal is a beautiful way to capture your feelings and responses.

Connect with the natural world

Connect with the natural world

 What you need

- A large notebook or sketchpad. (See the 'make a notebook' activity.)
- Pens and pencils. (You might want to use watercolours too.)
- Glue and tape.

 What you need to do

1. Spend some time in a natural space, looking and listening, touching (and, if possible, tasting!). What has caught your eye? What has captured your attention? Why?
2. Think about how you want to record your thoughts. You could write about what you've seen and find out the scientific names of the plants and animals you've seen. I love the Latin name for the bumblebee: Bombus! Maybe record the weather – how windy is it? What do the clouds look like?
3. You might want to attach beautiful leaves and petals from the ground and draw and colour some of the interesting plants you see. You could also make rubbings from trees and leaves. You could squash plants into the page (see the 'make paint from plants' activity).
4. It's good to go back to the same spot through the different seasons and record the changes you see.

 Extension activity

Create a map of the natural area and create your own key. This could be a map representing that day, showing where sticks and leaves are lying at that time.

Connect with the natural world

64 Grow potatoes

Why?

Reducing food miles helps the environment, and growing food helps connect us to nature. Plus food that you have grown yourself tastes so much better!

What you need

- Seed potatoes.
- Empty egg boxes.
- Potato bags.
- Compost.

What you need to do

1. Put the seed potatoes in the empty egg box on a sunny window sill. This is called 'chitting.'
2. Fill the potato bags with compost and put them in a warm sunny spot.
3. When small shoots appear on the potatoes, push them into the bag, with the shoots upwards.
4. Water the potatoes regularly.
5. When flowers appear or the leaves start to turn yellow, your potatoes are ready.
6. Tip up the bag and pull out all the potatoes.

Extension activities

- Other vegetables to grow include peas, beetroot and rainbow chard. You will need seeds, compost and pots. (See the newspaper pots activity.)
- Why not try no dig gardening? Put down a large piece of recycled cardboard. Cover this with compost and plant plants in this compost.

Connect with the natural world

65 Stone soup

Why?

Stone soup is an old fable about sharing which can be a great way to introduce soup making. We waste around 9.5 million tonnes of food each year and soup is a great way to prevent waste. As long as you have some onions and garlic, you can add any other vegetables that need to be used up.

The story

One day a traveller arrived in a village, and as he walked into the village, everyone looked at him suspiciously.

'May I please have a little food?' he asked. 'I am tired and hungry.'

'No,' said one villager, 'we don't have enough to eat ourselves. Go away.'

The stranger just smiled. He collected some sticks and started a fire. Then he took a cauldron and placed it on top of the fire. He filled the cauldron with water and took a stone out of his pocket. He lifted the stone up so that all the villagers could see it…then he dropped it in the water.

One little boy said, 'What are you cooking?'

The traveller replied, 'I'm making stone soup and when it is ready, you can all have some. Stone soup is always delicious.'

All of the villagers were interested now and came out to watch. 'Oh,' the traveller said loudly, 'I wish I had some cabbage. That would make my soup so tasty.' One of the villagers rushed off to her house and brought him a cabbage. The traveller cut it up and added it to the stone soup.

'Fantastic!' He said. 'One thing that makes stone soup even more tasty is onion and carrots.' One villager hurried into his house and came back with some onions and carrots for the soup.

The traveller tasted the stone soup. 'It is nearly ready! I just wish I had some salt and butter. It would taste even better then.' One woman went to her house and came back with salt and butter for the soup.

Now the traveller tasted the soup again. 'The only thing that could possibly improve this soup now is leeks and potatoes.' Everyone agreed and someone brought potatoes and leeks.

Connect with the natural world

Everyone could spell the delicious soup. 'I want some soup!' said a little girl. 'Is it ready yet?'

The traveller smiled. 'Soon you can have some. The soup just needs some herbs.' A villager gave him some herbs to go into the soup.

Finally the soup was ready and the traveller made sure everyone in the village had a bowl of it. Some of the villagers wanted to buy the magic stone but the traveller refused. He finished his bowl of soup, put the stone back into his pocket and continued with his journey.

 ## What you need

- Knives and chopping boards.
- Salt and pepper.
- Wooden spoons.
- Cups to drink the soup from.
- Oil.
- Water.
- A stock cube.
- Either a slow cooker or a pan for cooking over a fire.
- Optional: a stick blender.

 ## What you need to do

1. If you are using a slow cooker, cut up all the ingredients and add them to the slow cooker with water and a stock cube. Cook on high for four hours. If you chop the ingredients finely, it will cook more quickly.
2. If you are cooking over a fire, fry the onions and garlic. Add the chopped vegetables, water and a stock cube and cook until the vegetables are tender.
3. If you like, you can blitz the soup with a stick blender.

Tip from Poppy Hinton age 9 of the Forest Foxes:

'Preparation is key, ensure you have everything you need ready and accessible especially things you're going to need once inside the fire circle so you haven't got to keep moving away from the fire; things like your stirring spoon, soup ladle and the mugs for when it's ready.'

 ## Extension activity

Make croutons from slightly stale bread that is about to be thrown away. Cut the bread into small cubes and sprinkle with salt, pepper and olive oil. Bake in the oven on a baking tray until crispy.

Connect with the natural world

Connect with the natural world

Connect with the natural world

66 Make an upside-down herb garden

 Why?

We need to look at growing in different ways and herbs are one of the easiest types of plants to grow.

Connect with the natural world

What you need

- Clean, empty plastic bottles, with the labels removed.
- Scissors.
- Twine or string.
- Compost.
- Small herb plants or seeds. Herb plants that do not grow too big work best such as thyme, basil, oregano, marjoram, parsley, mint or coriander. Avoid plants like rosemary.
- Hole punch.
- Newspaper or tablecloths to avoid mess (or you could do this activity outside).

What you need to do

1. Cut the bottom off the bottle and dispose of this part. Be careful of sharp edges.
2. Use the hole punch to make three holes around the edge of the part you have just cut.
3. Gently push your small herb plant through the wider part of your bottle until the leaves are poking through the small opening.
4. Next, put some compost into the bottle.
5. Push string or twine through the holes you punched and tie them together so that you can hang your upside-down herb up.
6. Remember that when you water it, the water will drip down so consider that when you choose a place to hang it!

Extension activity

Plant herbs outside in a wooden pallet. Attach a piece of old tarpaulin or an old sheet to one side of a pallet. Turn the pallet over and fill it with compost. Plants herbs in the pallet. Water regularly.

Connect with the natural world

67 Making seed bombs and seed paper

🌍 Why?

Wildflower patches are welcoming and inviting to a wide range of creatures from butterflies to frogs, from dragonflies to birds.

🛠 What you need

- One tub of wildflower seeds.
- Compost.
- Flour.
- Water.
- Bowls for mixing.
- A tray.

What you need to do

1. Mix together the seeds with three tubs of compost and two tubs of flour. (Use your empty seed tub to measure.)
2. Keep adding little bits of water until you can form small seed balls that hold together.
3. Place the seed balls in a tray and leave them in a cool place for 24 hours. (If it is too warm, they may become mouldy.)
4. As long as you have permission, you can throw your seed balls in an outdoor space and wait for them to grow!

 ## Extension activity

Wrap the seed bombs in a piece of scrap fabric tied with string to make a gift. Make sure they are stored in a cool, dry place.

Make seed paper

 ### Why?

This is a great way to use up old paper and encourage wildflower planting.

 ### What you need

- Paper from the recycling bin such as newspaper or printing paper.
- A large bowl.
- Hot water.
- A stick blender.
- A tray.
- A dish cloth.
- Some wildflower seeds.
- A spoon.
- Cookie cutters.
- A tray.

 ### What you need to do

1. Rip the paper into small pieces and soak it in hot water for around 15 minutes.
2. Use the stick blender to blend the paper and water mixture. Check there are no big pieces of paper left.

3. Use your hands to squeeze water out of the mixture. Try to remove as much paper as possible.
4. Sprinkle the seeds into the mixture and mix with a spoon.
5. Put the cookie cutters onto a tray and put a little mixture in each one, pressing it down. Make sure it is not too thick.
6. Leave it to dry in a warm environment. (If you leave it to dry for too long, the seeds may start to grow! You can use a hairdryer on a low setting.)
7. When it is dry, store in a cool, dry place.

 Extension activity

Make your seed paper into a gift by sticking it to a piece of folded card to make a greeting card. You can even write someone's name on the seed paper with a permanent marker.

68 Sediment jars

 Why?

Sometimes we take soil for granted. It is incredible stuff but takes hundreds of years to form from rock. If you pick up a handful of soil, you could be holding more living organisms than the number of people on Earth! This includes things you can see without a magnifying glass like worms, ants and beetles, and also bacteria and algae.

Soil filters rainwater, stores nutrients for plants and helps prevent flooding. Unfortunately, at least a third of the soil on Earth is in poor condition because farmers have used too many fertilisers.

 What you need

- Clean, empty glass jars with lids.
- Trowels or small digging tools.
- Soil.
- Water.

Connect with the natural world

 ## What you need to do

1. Half fill your jar with soil. It's a good idea to collect in different areas so that you can compare the soil in different zones. Take out any visible living organisms.
2. Add water to the jar almost to the top. Put the lid on and shake it or stir the contents with a stick.
3. Now put the jar somewhere safe and leave it to stand undisturbed for at least one day.
4. Have a look at the layers when the soil and water have settled. What can you see? Can you draw the different layers? You may see organic matter, clay, silt, sand and gravel.

 ## Extension activities

Measure the total height of the total layers and then each layer. Work out the percentage of each layer and compare it with soil from other areas. Find out if the soil is acid or alkaline. If you add half a cup of vinegar to a handful of soil and it fizzes, it is alkaline. If you add half a cup of baking soda to a handful of soil plus a little water and it fizzes, it is acidic.

Connect with the natural world

69 Natural paintbrushes

Why?

This helps us explore the textures of different natural materials and create art using nature.

What you need

- Sticks.
- Natural materials: collect different plants that will make interesting patterns. You could try grass, leaves, feathers, ferns.
- Elastic bands or string.
- Paper.
- Pots of paint.

Connect with the natural world

 ### What you need to do

1. Attach the natural material to your stick with an elastic band to make a 'paintbrush.'
2. Dip your new paintbrush into paint and experiment with different ways to make marks on the paper.

 ### Extension activity

You could try sweeping the brush across the paper or dabbing or swirling. What type of picture can you create?

Connect with the natural world

70 Painting with plants

🌍 *Why?*

This activity boosts nature connection, cultivates creativity and provides opportunities to explore cause and effect.

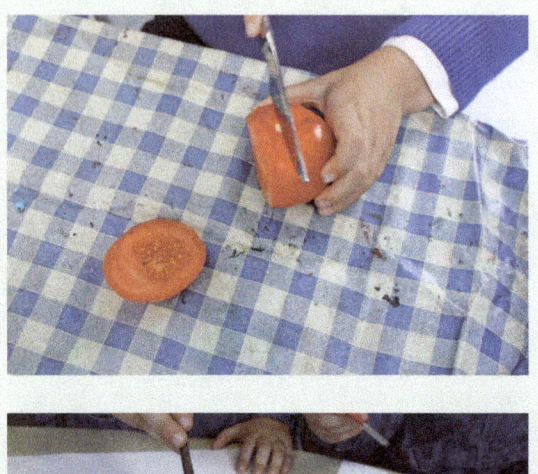

Connect with the natural world

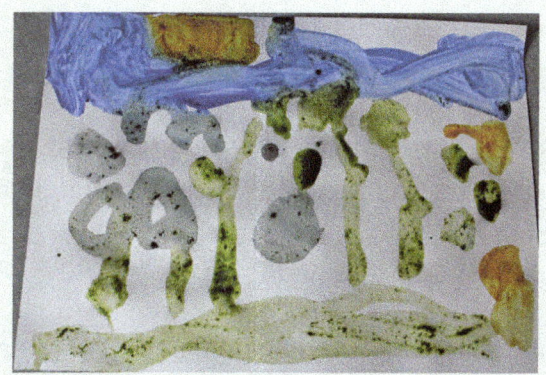 

What you need

- Flowers and plants. Dandelions, mustard powder and turmeric make yellow paint. Fresh spinach makes green, blackberries make blue, and beets, blueberries and red cabbage make red. If you add baking soda to the red cabbage, it turns blue. When you add vinegar it turns pink. Pickled beetroot from a jar is easier to use than fresh beetroot as it is already cooked.
- Knives and chopping boards.
- Aprons/old shirts.
- Thick paper for painting.
- Paintbrushes.
- Small jars to hold your homemade paint.
- A smoothie blender or similar.
- Flour.

What you need to do

1. Talk about why flowers are generally bright colours. (These bright colours attract insects. Bees are most attracted to purple, violet and blue, but they also like yellow dandelions.)
2. With your aprons on, chop and squish each different plant. Then, with an adult helping, blitz the small pieces in a blender, adding water gradually and a little flour until it feels like paint. (Wash the container out after each different plant.)
3. Pour your paint into containers and paint! What differences do you notice between this homemade paint and the paint you usually use?

Ledbury School: 'The blueberry paint smells yummy! I'd like to eat it! It's a bit thinner when it dries and it fades.'

Connect with the natural world

 Extension activity

The next stage could be to dye material with your plants. If you are using red cabbage, spinach or dandelions, chop them up into small pieces and add twice as much water plus 250ml of vinegar. Heat in a slow cooker for about an hour and leave it to cool. Blitz it in a blender. If using berries, substitute salt for the vinegar.

You could tie-dye some material. Dip the material in water to make it damp. Gather up small sections of the material and tie them up with elastic bands in random places. Put the material in a container and pour the dye over it. Put the container in a plastic bag and leave it in a warm place for 24 hours. Cut off the rubber bands and rinse the material in cold water. Leave it to dry.

Connect with the natural world

71 Nature scavenger hunt

 Why?

This helps us to really get to know a natural space.

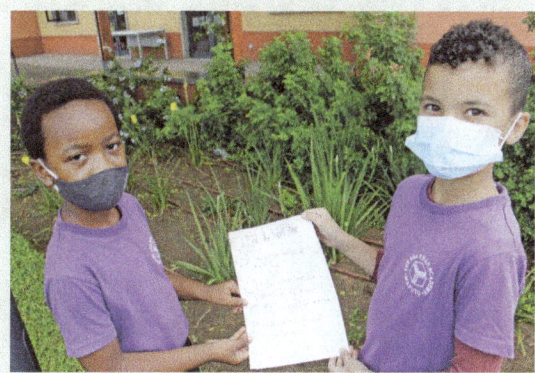

Connect with the natural world

 What you need

- Access to a natural space.

 What you need to do

A scavenger hunt is a game where someone challenges the group to find certain items. You could set a time limit and you could work in teams.

Here are some ideas for your nature scavenger hunt:

- Have a colour scavenger hunt: 'Can you find something green? Something yellow? Something brown?' You could use the paint sample sheets from DIY stores and find items that match.
- Find different textures: 'Can you find something smooth? Something prickly? Something slimy?'
- You could ask the group to find something they think is: beautiful, precious, thin, thick, round, alive (be gentle!), noisy, bright, dark.
- An adult could take some photos of plants and creatures in the natural environment and everyone could try to tick all of these off.

Aga Khan Academy, Mozambique:

Kiran (age 7): 'To make it more exciting add details: instead of a leaf you could look for a *pointy* leaf; instead of a flower hunt for a purple flower.'

Taissa (age 10): 'Choose some things that only come out in one season – then you can talk to younger children about different seasons and how nature changes.'

Tino (age 6): 'Make some rules before you go, like avoid treading on flowers and always look around to make sure you are safe.'

72 Carry out a bio blitz

🌍 Why?

A bio blitz shows us which plants, animals and other living things are in one spot. It helps us really get to know a natural area.

Connect with the natural world

 What you need

- Magnifying glasses.
- A clipboard and a sheet of paper for recording.
- Pencils.
- Some hula hoops or string tied together to make a circle.

 What you need to do

1. Get into small groups and find an area to work on. You may want to compare very different areas: for example, an area of long grass and a tarmac area. Decide where you are going to look – will you look underground too? Decide how long you will look for and set a timer.
2. Put down your hula hoop on the ground and count and record all the creatures inside the hoop.

 Extension activity

Use an identification app to identify the creatures you find.

73 Hapa zome

 Why?

Hapa zome means 'leaf dye.' It's smashing leaves or other parts of plants onto fabric to make a pattern. Hapa zome gets us hands-on with nature and helps us see the beauty of plants. All those lovely colours you see outside in the summer bring important pollinators over like bees.

 What you need

- Pale coloured fabric. Cotton works best but you could try different types of material.
- Flowers and leaves. Try to take them from places where there are a lot of those leaves and flowers so that there is enough for minibeasts and other creatures.
- A hammer, mallet or hard object for smashing the leaves and flowers.
- A hard surface like a wooden chopping board or table to do your smashing on. This surface may get stained by the flowers and leaves so make sure you protect it with a wipeable tablecloth if necessary.

 What you need to do

1. Cut your fabric. Think about how you will display your finished artwork. You might want to make triangle bunting or you might want to work on square pieces that can be framed. However you decide to cut up your fabric, remember that you need a 'spare' piece of fabric to place over the top while hammering.
2. Collect the flowers and leaves you want to use.
3. Place a piece of fabric on your hard surface and arrange the flowers and leaves on top. You can create a design or arrange them randomly and see how they come out.
4. Put your spare piece of fabric on top.
5. Hold the fabric steady with one hand and tap the fabric with a hammer or similar object, taking care not to hit your hand! You will see straight away if it is working, because the flowers and leaves will start to stain the material.
6. When you have smashed all the flowers and leaves, take off your spare piece of fabric and peel off the flowers and leaves. Admire your artwork!

Connect with the natural world

 Extension activity

Make a picture frame for your creation using recycled cardboard.

Emma Curthoys, Headteacher, Solent Infant School: 'The children explored hitting and also rolling with pressure. They loved it and worked out that the fresher the leaves, the better the imprint.'

Connect with the natural world

Connect with the natural world

74 Land art

Why?

Land art helps you use your imagination as well as connecting with nature.

What you need

- Natural materials like leaves, sticks, pine cones, shells, stones etc.

What you need to do

Anyone can make land art (making art and sculpture from natural materials) and the outdoors is ideal for land art.

Why not look at some artwork by Andy Goldsworthy and Ugo Rondinone?

Arrange your natural materials in a pattern or a picture.

Connect with the natural world

Connect with the natural world

Connect with the natural world

75 Make a nature mobile

 Why?

This is an interesting way to display natural materials from your local area. Maybe you could go on a walk to find these materials...

 What you need

- Sticks.
- String, wire or wool.
- Scissors.
- Natural materials.

 What you need to do

1. Choose a selection of natural materials like leaves, pine cones, feathers, small twigs.
2. Attach these to lengths of wool, string or wire by piecing them and threading them or tying them on.
3. Attach the string, wire or wool sections to your stick.

Top tips from Mylnhurst School:

1. If you are using pine cones as part of your nature mobile, dry these out in front of a warm fire. This will make the pinecones open up!
2. Wear gardening gloves when collecting prickly items, such as holly.
3. We are going to make nature mobiles again in spring, summer and autumn – it will be interesting to see what different natural materials we can find in the different seasons.
4. We used card making cutters to punch out decorative shapes in the leaves to add extra detail.

 Extension activity

Make a journey stick. Take a walk through a natural area and select natural objects as you walk along. Attach each item to a stick with string or wool.

Connect with the natural world

Connect with the natural world

Connect with the natural world

76 Pine cone activities

Project 1: Pine cone bird feeder

 Why?

Birds need food that is high in fat in the winter when food is scarce.

 What you need

- Pine cones.
- Lard or suet (at room temperature).
- Birdseed.
- A mixing bowl and spoon.
- String or twine.
- Scissors.
- A tray and baking paper.
- A fridge.

 What you need to do

1. Tie a length of string around the top of each pine cone.
2. Mix your lard or suet with the birdseed in the bowl.
3. Take some of the mix and pack it tightly around the pine cone so that you have a ball shape.
4. Put your pine cone feeders on baking paper on the tray in the fridge for a few hours.
5. Hang your bird feeder outside in a place where birds can safely feed.

Connect with the natural world

Project 2: Pine cone garland

 Why?

Fallen pine cones are easy to find in autumn and this is a great way to display their beauty.

Connect with the natural world

 What you need

- Pine cones (don't take too many from one spot as some wild animals eat them).
- String or ribbon.
- A hot glue gun (alternatively you can simply tie them on).
- Scissors.

 What you need to do

1. Cut a long length of string or ribbon.
2. Cut some shorter lengths of string or ribbon.
3. Attach a pine cone to each shorter length, using a glue gun.
4. Tie these shorter lengths to the long lengths at regular intervals.

77 Nature prints in clay

 Why?

Making clay impressions of natural objects gets us to look more closely at the nature around us. We start to really notice the wonderful details in plants and even weeds. Making prints helps us see nature in a different way: we begin to think about the texture of a plant and the impression it will leave.

Connect with the natural world

Clay is a wonderfully squidgy natural material which can strengthen our hands and wrists. Working with clay can also be good for our mental health because it reduces our worries and makes us feel calm.

 ## What you need

- Air drying clay. (Or you can make your own salt dough.)
- Natural materials like flowers, herbs, ferns, pine cones, feathers. In the winter, search for evergreen plants such as fir trees. Strong leaves with visible veins are best.
- Tablecloths to protect surfaces.
- Tools for cutting, rolling and shaping clay.
- Rolling pins.
- Baking paper or tracing paper.
- Cookie cutters. Circular cutters are best. Fiddly shapes like stars can be tricky.
- Pencils.
- Ribbon.

 ## What you need to do

1. Collect the natural materials you want to use to make impressions. You could always practise on playdough if you want to see what the print will look like. It can be a nice idea to go on a nature journey nearby because then all the materials you collect will remind you of your walk. Try to avoid very fragile plants which will break easily when you push them into the clay.
2. Make sure everyone has a good-sized ball of clay. Put your clay ball onto a piece of baking paper. Next, roll your clay out into a small pancake shape. Don't roll it too thin or your natural materials will go right through.
3. Press your natural materials down onto the clay and run a rolling pin over the top. Peel your natural materials out and see how your print has come out. If you are happy with it, put a cookie cutter over your favourite section and cut it out. Use the non-writing end of a pencil to make a small hole for your ribbon. Let it dry. (It should take about 24 hours.) Thread a small piece of ribbon through the hole.

 ## Extension activity

You could paint your dried clay impression before threading the ribbon through.

Connect with the natural world

78 Stick fairies/ stick characters

 Why?

This activity gets us to examine natural materials carefully and see their potential.

Copyright material from Sarah Watkins (2023), *99 Eco-Activities for Your Primary School*, Routledge

Connect with the natural world

 ### What you need

- Sticks.
- Leaves and other natural items.
- Felt tip pens/paints.
- PVA glue.

Optional:

Knives or peelers for whittling.

Small saw for cutting sticks to size.

Air drying clay to build up your character.

 ### What you need to do

1. Choose a stick that you feel will work well.
2. Snap it to the size you want.
3. You might want to whittle it to provide a good surface for painting.
4. Decorate the stick to make your character.
5. Glue on leaves and other items.

Hayley Reading, St. Peter's Catholic Primary School, Marlow:

'The children really loved this activity!

Top tips:

- When working with lots of children at a time, it helps to pre-whittle the sticks. We just whittled enough space for the faces as it kept the lovely textures we had of all of the different sticks around school.
- We used leaves that we found close by but it would be nice to go on a leaf hunt to find different sized and shaped leaves that could give the character different "looks."'

 ### Extension activity

Could you also make a home for your character?

Connect with the natural world

79 Minibeast watching

 What you need

- A clean plastic container or jar with a lid to observe bugs more closely.
- A magnifying glass.
- Access to a natural space.

Connect with the natural world

 ### *What you need to do*

1. In the spring and summer, you will find bugs on flowers and other plants. You will also find insects and other minibeasts in hidden places such as under stones and logs.
2. Use your magnifying glass to look more closely at them and you could gently place them in your container if you want to see them close up.
3. You could draw them.
4. Just remember not to keep them in the container for too long and to release them back where you found them.

 ### *Extension activity*

- ID the creatures you find using an app or a book.
- Make your own pooter. A pooter is a container that helps you to look at bugs closely. Take a small plastic container with a lid from the recycling bin. Cut a hole on one side and a hole on the opposite side, big enough to push a paper straw through. Cut a small section out of some used, clean tights and attach this over the end of one straw with clear tape. Push this end of the straw into one hole and push the end of another straw into the opposite hole. Secure these straws in place with blue tack. Put the lid on the container. Take your pooter outside and look for small bugs. Using the tights covered straw, suck insects up and observe them for a short time. Don't try to suck up insects larger than the straw.

80 Build a minibeast hotel

 Why?

Minibeasts are invertebrates (creatures without backbones). For example, spiders, slugs, beetles and worms. They are vital to our ecosystem – our lives would be almost impossible without them. Minibeasts pollinate around a third of our crops and they decompose dead leaves and other materials. They are also a vital part of the food chain. If global warming continues, we risk losing many species of insects.

Connect with the natural world

By building a minibeast hotel, we can provide a safe place for minibeasts to live and we can observe them, building a connection with them.

What you need

- For a 'takeaway' bug hotel: clean plastic drinks bottles or milk bottles. Twigs, pine cones etc.
- Recycled materials such as pallets, broken terracotta plant pots, small plastic plant pots, roofing tiles, sections of pipe, bricks, chicken wire, toilet rolls.
- Natural materials such as twigs, leaves, bamboo canes, pine cones, bark, logs, plant stems, straw, nutshells, stones, pebbles, moss, dead wood.

What you need to do

1. If you want to make bug hotels that you can take away, you can cut the middle section from a clean, used plastic bottle, and fill it with sticks, leaves and other materials.
2. For a larger bug hotel, make sure you have a partly shady spot and that the bug hotel is built on a solid base.
3. Stack your pallets in layers, with spaces between the layers. You can add bricks to make the layers bigger. Now fill the spaces with twigs, leaves, sections of bamboo cane, pine cones, bark, straw, stones and pebbles.
4. Add a roof to keep most of the rain off.
5. Plant some flowers nearby to attract bees.

Connect with the natural world

81 Keeping chickens

 Why?

Keeping chickens reminds us where our food comes from and helps us to learn how to care for animals. You will also get a daily supply of fresh eggs!

Chickens have wonderful characters and can be great fun. Chickens are not too much hard work but you do have to look after them every day and there is a cost. It's a good idea to talk to an organisation like https://www.hensforhire.co.uk/ to find out whether it will work for you.

You could work with a local farm that can lend you some chickens plus all the equipment.

 What you need

- Chickens! Think about which breed will be best. Pure breeds can be more docile.
- A chicken house and a run.
- A feeder and drinker.
- Bedding.
- Chicken food.

 What you need to do

1. Remember to wash your hands after touching the chickens.
2. Let the chickens out of their house into their run every morning.
3. Chicken need food and water every day so you do need to think about weekends and holidays.
4. Check for eggs every day.
5. Tidy their house and run every day and give it a good clean every week.

Connect with the natural world

82 Ice bird feeder

 Why?

Birds can be short of food at any time of year and watching them using a bird feeder is exciting! We have a responsibility to help other living creatures.

Connect with the natural world

 What you need

- Water.
- Birdseed mix/sunflower seeds/millet/apple pieces. You could add some cranberries – these are safe to eat but not all birds enjoy the sour taste.
- Used, clean plastic food containers without holes like yoghurt pots or small margarine containers. (Or ice cube trays to make small feeders.)
- String or twine.
- Scissors.

 What you need to do

1. Partly fill the containers with water.
2. Pour some bird food into the containers.
3. Cut short lengths of string or twine and put one end in the container, making sure a good section of it is in the water.
4. If the temperature is going to be very cold, leave the containers outside overnight. If not, put them in the freezer.
5. When you are ready, push the iced feeders out of the containers and hang them up outside. Think about the best location – it's better if it is in a quiet place where birds can see predators like cats approaching.

 Extension activity

- Make an apple bird feeder by removing the apple core and putting string through the hollowed centre.
- Use hoop-shaped cereal and pipe cleaners to make a quick bird feeder to take home.
- Use clean, empty juice cartons to make bird feeders.
- Make your own bird food and put in paper bags to sell to raise money for your school or setting or for an eco cause.

Connect with the natural world

83 Make your own bird's nest

🌍 Why?

Birds are a vital part of our ecosystem because they control pests and distribute seeds. Some birds like the curlew and turtle dove are on the RSPB's list of threatened birds. This is a good way to find out how hard it is to make a nest!

Connect with the natural world

What you need

- Sticks and natural materials such as moss.
- Fresh eggs.

What you need to do

The challenge is to make a nest suitable for a bird and this is not easy! When nest building, birds drop some small sticks to make a base and then weave other sticks together with their beak to make the walls of the nest. Adding moss and other natural materials makes the nest cosy. They also use spit and spider webs to stick the different elements together.

1. Collect natural materials, thinking about which will work best.
2. Try weaving bendy twigs.
3. Can you add some soft natural materials?
4. When your nest is ready, give it a blow – does it fall apart?
5. Put an egg gently into the nest. Does the nest support the egg?

Drapers' Pyrgo Priory School: 'We did this activity with Reception and the children are just beginning Forest School, so the fine motor aspect of this activity was quite tricky. We tried to use materials that were found rather than growing. I hard-boiled the eggs in case they rolled out of the nests. I introduced the activity using a plush bird. We had lots of fun doing it!'

Extension activity

- Watch birds using binoculars.
- You could even make a birdwatching hide using branches and sticks.

Connect with the natural world

84 Make a bird bath

 Why?

Hopefully you have a wash most days! Birds also like to have a bath. They need to keep their feathers in good condition.

Connect with the natural world

 ## What you need

For a takeaway bird bath to use at home: plant saucers or shallow recycled food containers without holes, small stones and gravel.

For a larger bird bath:

A large terracotta pot and a large plant saucer that will sit on top of the upturned flower pot.

Or…

A large plant saucer, dustbin lid or shallow plastic tray. Three or four bricks to raise it off the ground.

 ## What you need to do

1. To make a takeaway bird bath, use a recycled plastic food container or plant saucer. It must be shallow so it is safe for birds. Put some gravel or little stones into the container. When you get it home, put it outside, somewhere high and fill it with water.
2. To make a larger bird bath, decorate a large terracotta pot with paint. Find a good spot for the bird bath. It's best if you put it in an area that is not too busy and where birds can see predators like cats coming.
3. When the paint is dry, put the pot upside down in a good spot. Put the plant saucer on top and put some gravel inside. Pour some water on top of the gravel and wait for birds to come and use it!
4. Or place bricks on the ground where birds can see cats coming. Put your large saucer, dustbin lid or tray on top and make sure it doesn't wobble. Put in some gravel and small stones to stop the birds from slipping.

85 Butterfly feeder

 Why?

Butterflies are pollinators and their sources of food are being reduced. This butterfly feeder gives them the chance to stop and feed.

Connect with the natural world

What you need

- Clean empty food containers like fruit containers.
- Hole punch.
- Scissors
- String.
- Fruit. Butterflies love overripe bananas, strawberries, pears, plums and apples.

What you need to do

1. Cut the container so it is nice and shallow.
2. You might want to paint some bright colours onto the container to attract butterflies.
3. Punch holes at each corner of the container.
4. Push string through the holes and tie the string together so that the feeder can be hung up.
5. Cut your fruit up into small pieces and place it in your feeder.
6. Hang the feeder up outside.

Extension activity

Put out a plate of sliced up overripe bananas and halved oranges to attract butterflies.

Connect with the natural world

86 Make a hedgehog home

 Why?

Hedgehogs are in decline and are listed as 'vulnerable.' They have less nesting spots and many of their habitats have been destroyed.

Connect with the natural world

 ### *What you need*

- A piece of old wood like plywood.
- Some old unwanted bricks.
- Fallen leaves.

 ### *What you need to do*

1. Find a space that is sheltered and quiet.
2. Build your hedgehog home by laying the bricks in a rectangle, leaving a small gap for the hedgehog to get in. The hedgehog home only needs to be around three bricks high.
3. Now build a short tunnel (one brick length) by the door.
4. Put some leaves inside the hedgehog home and then put the wood on top as a roof.
5. You might want to place some more bricks or a few old logs on top.

 ### *Extension activity*

- You can simply put a paving slab on top of four bricks.
- Another idea is to use an old plastic storage box. Cut an entrance into the box, put some old leaves and twigs inside, then cover the box with twigs, leaves and natural materials.

Connect with the natural world

87 Mud faces

 Why?

Mud is wonderful! It feels great in your hands and it has bacteria inside that makes you feel happy.

Connect with the natural world

What you need

- Mud! (You can also use air drying clay.)
- Natural materials for decorating.

What you need to do

1. You may need to add a little water to your dirt to make mud.
2. Shape and sculpt the mud until you have a face shape.
3. Use natural materials to make eyes, nose, mouth and other parts of the face.
4. You could make your mud face on the ground or stick to a tree or make a face that sticks onto a stick.
5. Leave the mud to dry.

Extension activity

Could you do some cooking with mud? Can you write your recipes down and illustrate them?

Connect with the natural world

88 Whittling

This requires close adult supervision.

 ## Why?

Whittling helps you feel calm and helps you concentrate. Using real tools is a good skill.

 ## What you need

- Sticks. (Avoid yew and laurel, which are toxic.)
- Potato peelers.
- Vegetables like carrots.
- Knives. Using a blunt knife can be dangerous. It is best to use knives sold by Forest School suppliers. These must be used with adults guiding.
- Gloves made from tough fabric.
- First aid kit in case of accidents.

Connect with the natural world

 ## *What you need to do*

1. The adult needs to model how to use the peeler. Everyone can start with a vegetable such as a carrot, to get used to using the peeler. Then the adult can demonstrate how to strip bark from a stick, wearing a glove on the hand they don't write with. Push the peeler away from your body. Make sure there is plenty of space between you and another person.
2. It works well to sit down and hold the stick in the hand you don't write with. Put this hand, holding the stick, over your knees, away from your body. Use the peeler to remove as much bark as possible, taking off small strips at a time.
3. Next, the adult needs to introduce the knives. These should be kept in a locked box when not being used and an adult must supervise a child whittling with a knife. You should have your thumb on the blunt side of the blade, with the rest of your fingers firmly gripping the handle.
4. Now try some whittling projects!

Connect with the natural world

89 Build a den

 Why?

Some people around the world don't have proper shelter because of earthquakes, floods and other natural disasters caused by climate change. Building dens helps us to see how hard other people have to work and makes us think about what materials people need to use to be safe.

Connect with the natural world

By making a den outside from natural materials, we start to see how different materials can be combined to make the best den.

What you need

- If you are making dens indoors, you will need things like blankets, den clips or pegs, chairs, large cardboard boxes.
- If you are making dens outside, you will need things like tarpaulins, long sticks or poles, clips, string.

What you need to do

1. Do you know which animals like to live in a den? Animals like rabbits, foxes and polar bears live in dens!
2. Decide what the den should be like: will it protect you from rain? Will it be warm and cosy? Can you decide on a test for the den? For example, water could be poured onto each den from a watering can.
3. Start to build the dens, changing your plans if the den doesn't work. Now, test the dens! How did they do?

90 Take one stick

Why?

Getting creative with a stick helps us to build a nature connection. It encourages us to take notice of the nature all around us.

What you need

- Some sticks!

What you need to do

Get creative with your sticks. Here are some ideas to get you started:

Make a wand.
Make a journey stick.
Make a flag to represent your school or setting or your family.
Make a God's eye.
Make a mini tipi.
Make a stick butterfly.
Make arrows.
Make pixie sticks: use a peeler to sharpen the end of a stick then collect as many leaves as you can on the stick.

Connect with the natural world

Connect with the natural world

91 Slack line

 ### Why?

Slacklining helps improve balance and core strength. It requires lots of concentration! Moving differently in nature helps us to connect with nature.

 ### What you need

- You need a slackline set that will attach to trees.
- You also need two trees! It's best to use tree protectors so that the slackline does not damage the tree.

 ### What you need to do

1. Follow the instructions that come with your slackline and attach each end to a tree. Make sure the bottom slackline is low to the ground so that it is easy to get on and off. If you fall off, you won't get injured! Make sure it is as tight as possible – it will bounce and stretch!
2. It's best to go barefoot. Practise balancing on one foot then try to walk across the slackline.
3. Once you feel confident, try walking backwards, forwards, sideways and bouncing!

Connect with the natural world

92 Climb a tree!

 Why?

Tree climbing helps you stay flexible and improves problem-solving skills too. Climbing a tree helps you see nature from a different angle.

Connect with the natural world

 ## *What you need*

A tree! Trees with low branches work best. Get to know your tree first. Look at it from every angle. If it is a very young tree, it may not support your weight.

 ## *What you need to do*

1. Don't climb trees in a thunderstorm or strong winds. In very cold weather, watch out for icy branches. Never climb a tree that is close to a power line.
2. Your adult might set some rules. One rule might be how high you are allowed to climb. It's important that everyone follows the rules. Take turns: having more than one person in a tree can get dangerous.
3. When you are moving along a branch, test it gently with your foot before putting all your weight onto it. Think carefully about where your feet are going to go next. When you move your hand to a different branch, tug it first with your hand before putting all your weight onto it. When you come down, do it slowly and ask for help if you need it.

Comments from the children at Earthtime for All:

"I can reach the next branch easier now I'm bigger."

"I like to sit on this branch."

Zoe Sills of Earthtime for All: "Our rope ladder is a great resource that helps to develop the strength and confidence to translate to climbing trees."

Connect with the natural world

93 Cloud spotting

 Why?

A cloudy sky changes all the time and if we take some time to look at clouds, it can make us feel happier!

Connect with the natural world

 What you need

- A sky with a few clouds but not completely covered by clouds.

 What you need to do

Lie on your back (put down a blanket or a tarpaulin if it is damp) and look up at the clouds. What can you see? Describe the shapes you can see. Are they changing?

 Extension activity

- Make shadow art. On a sunny day, put out long rolls of paper and place objects down to make shadows. Draw around these to create art.
- You could also use chalk or powder paint to colour in outdoor shadows.

Connect with the natural world

94 Lay a trail

🌍 *Why?*

This activity gets us to look closely at a natural area and notice what might be out of place.

Connect with the natural world

🛠 What you need

- A natural space.
- Some 'treasure.' This could be a shell or a distinctive stone.

🌱 What you need to do

1. Split into two groups.
2. One group needs to look away while the other group lays a trail.
3. The trail could include sticks on the ground in the shape of arrows or pine cones in a shape. The 'treasure' can be hidden at the end of the trail.
4. Each group follows the trail and tries to find the 'treasure'!

💧 Extension activity

Make an orienteering map and then make some markers from recycled materials. You could make flags from sticks and old cardboard from cereal boxes. Draw a map of the area which shows where other children can find the markers.

Connect with the natural world

95 Make natural noughts and crosses

 Why?

This is a great chance to test your game playing skills while also using natural objects.

 What you need

- Twigs and sticks.
- Stones or pebbles.
- String.

 What you need to do

1. Choose four sticks and make a noughts and crosses grid.
2. Find some small twigs which will fit in each gap.
3. Take two twigs and tie them together with string in a small cross shape.
4. Find some small stones that will fit in each gap.
5. Play noughts and crosses!

 Extension activity

Can you make your own game, using natural materials?

Connect with the natural world

96 Plant trees

 Why?

A tree can remove one tonne of carbon dioxide from the air every year. Trees provide shade and they help cool down the earth around them.

 What you need

- Saplings. Fruit trees work well in most spaces. Alder, silver birch and hazel also work well.
- A spade.
- Watering can.

 What you need to do

1. Think carefully about the best place for the tree(s) to go. Make sure you have permission to plant trees in that area.
2. Soak the tree in water.
3. Dig a large hole in the ground.
4. Loosen the roots of the tree to encourage the roots to grow into the soil.
5. Carefully place the tree into the hole.
6. Push the soil back into the hole so that the tree is covered.
7. Water the tree regularly.

Connect with the natural world

97 Make a sound map

Why?

This helps us to be in a natural space in a different way. This task makes us use our ears rather than our eyes.

What you need

- Pieces of recycled cardboard.
- Scissors.
- Marker pens.

What you need to do

1. Cut a big shape out of cardboard.
2. Take your cardboard and a pen outside.
3. Walk around the space, listening carefully.
4. When you hear a sound, think about what picture you could draw to show that sound. Every time you hear another sound, draw it.
5. Get together and compare – did you hear the same sounds?
6. Which sounds are man-made sounds?

Extension activity

Display your sound maps as a sound gallery.

Connect with the natural world

98 Outdoor mindfulness

Project 1: Sit spot

 Why?

Spending time in a special outdoor sit spot means that we really get to know this little area of nature.

 What you need

- Access to a natural space.

 What you need to do

- Find a special spot. This might be on a patch of grass, or on a log or on a bench.
- Get comfortable and sit there without talking.
- Think about what you can see. Think about what you can feel. What can you hear? What can you smell? Don't speak yet.
- Close your eyes and breathe deeply.
- When you have all finished, talk to each other about your special spot.

Project 2: Get to know a stone

 Why?

Getting to know a stone makes us look carefully at natural objects and really take in all of the details.

Connect with the natural world

 ### What you need

A selection of small stones of different colours, textures and sizes. (Enough for each person to have one.)

 ### What you need to do

1. This activity can take place indoors or outdoors.
2. Lay out all of the stones. Everyone should pick one. Take your time choosing one.
3. Take your stone and sit down. Look carefully at your stone. What can you see? Turn it over and explore each side. Run your fingers over it and touch every side of it.
4. Now everyone puts all the stones back. Everyone will turn away while someone muddles the stones up.
5. Now try to find your stone. How did you know it was your stone?

 ### Extension activity

Draw your stone.

Project 3: Stone balancing

 ### Why?

Stone balancing helps us concentrate and feel calm.

Connect with the natural world

 What you need

- Lots of different stones and pebbles of different sizes.

 What you need to do

1. Find a flat area to do your stacking like a tree stump or the playground or a flat piece of dirt. It's better to do stone stacking at your school or setting than in the wild. Otherwise it can disturb animals' habitats.
2. A large, flat stone needs to go at the bottom of your stack. Then choose other stones to make a tower. Take your time and don't rush!

 Extension activity

- Can you make an arch?
- Can you balance a stone on its end?
- Try balancing tiny pebbles between larger stones.

Project 4: Stone painting

 Why?

This is a calming activity that helps us to be creative.

 What you need

- Smooth stones.
- Acrylic paint.

 What you need to do

- Decide how you want to paint your stone and paint it!
- You can keep it, give it as a gift or leave it somewhere for others to find.

 Extension activity

Make stone photo holders

Take your painted stone and wrap a piece of wire around it. Leave a bit of wire at the top and wrap it round a pencil a few times – now you can push a photo into the loops!

Connect with the natural world

Project 5: Meet a tree

 Why?

Trees are the biggest plants on our planet and they are the longest living species on earth. We need trees because they release oxygen and store carbon, as well as providing food and shelter for animals and birds.

 What you need

- Some trees!
- Blindfolds/scarves.

What you need to do

This activity works best with pairs.

1. One person from each pair should wear a blindfold. Their partner guides them to a tree, looking out for obstacles that could be tripped over. To make it more tricky, you could guide your partner the 'wrong' way first and then the 'right' way!
2. The blindfolded person 'meets' the tree. Put your arms around it, feel the bark, reach up as high as you can and as low as you can.
3. Everyone goes back to where they started and the blindfolds are removed. Those children who had blindfolds now try to find their tree. Can they spot it?
4. Now swap over.

 ## *Extension activity*

Can you try to identify different trees in the area?

Project 6: Sit by a fire

 ## *Why?*

Sitting by a fire is fun but also helps us feel calm. It can help us bond with our friends.

This is an activity that must be supervised by adults.

 ## *What you need*

- A firepit area. If you do not already have one, an adult can mark out a firepit with a circle of stones where the fire will be and a circle of logs for people to sit on. If you have a small space, you can use a portable firepit and sit on mats around the fire.
- Tinder (twigs, wood shavings, dry leaves to start the fire).
- Kindling: small sticks.

- Used newspaper or paper to be recycled.
- Wood.
- Matches, a lighter or a flint.
- Petroleum jelly.
- Cotton wool balls.
- A fire blanket.
- A watering can full of water to put the water out.
- First aid kit.
- Kneeling mats.
- Optional: marshmallows or other food to be cooked over the fire.

What you need to do

This activity must be carried out with adults.

1. Make sure everyone understands that people must not walk across the fire circle. They must step over the log circle and sit down carefully.
2. Put a little petroleum jelly on the cotton wool balls and put them in the centre of the firepit. Scrunch up paper and put the paper balls around the cotton wool.
3. Take your twigs and place them standing upright around the paper, like a tipi.
4. Place a few small sticks around the twigs, standing upright. You don't want too many at first.
5. Light the cotton wool balls and, if necessary, get down and blow them to ensure that the kindling and tinder catch light.
6. As the fire gets going, add more small sticks and then a log.
7. Enjoy cooking or simply enjoy togetherness around the fire.
8. When you want to finish, pour water over the fire until it is out and leave this area for a few hours, because the embers may still be hot.

Extension activity

Why not cook damper bread over the fire? Rub together 80g butter and 500g of flour until it resembles crumbs. Mix in a teaspoon of salt, a teaspoon of sugar and a little water. (Keep adding water from 200ml until you have a dough.) Roll your dough into a snake shape and wrap it around a kebab stick. Hold it over the fire until it is cooked.

Connect with the natural world

99 Go barefoot

🌍 *Why?*

Going barefoot can be great for our health and it can make the muscles in our feet stronger.

Copyright material from Sarah Watkins (2023), *99 Eco-Activities for Your Primary School*, Routledge

Connect with the natural world

 What you need

- An area of grass that is safe from litter and rubbish.

 What you need to do

Check where you are treading and check for natural obstacles such as sticks and stones. Take a no shoes and socks journey!

 Extension activity

When you get more confident, try walking barefoot on different surfaces and through water.

CONCLUSION

I hope you have enjoyed these 99 eco-activities and that you are able to revisit them in the future and even adapt them. It's clear that we need to take action right now to protect our unique planet and ensure that future generations of children have access to clean air, food, water and shelter. But we also need to believe that we can all make a difference: we are all potential change makers. Climate change is a frightening reality but by taking action we can reduce fear and anxiety. Even by simply dipping into this book, you have joined a community of like-minded people, a community dedicated to positive action.

For Product Safety Concerns and Information please contact our EU representative GPSR@taylorandfrancis.com
Taylor & Francis Verlag GmbH, Kaufingerstraße 24, 80331 München, Germany

www.ingramcontent.com/pod-product-compliance
Lightning Source LLC
Chambersburg PA
CBHW082014220426
43670CB00015B/2623